Learn, Practice, Succeed

Eureka Math®
Grade 6
Module 2

Published by Great Minds®.

Copyright © 2019 Great Minds®.

Printed in the U.S.A.

This book may be purchased from the publisher at eureka-math.org.

4   5   6   7   8   9   10   LSC   26   25   24   23   22   21

ISBN 978-1-64054-965-4

G6-M2-LPS-05.2019

Students, families, and educators:

Thank you for being part of the *Eureka Math®* community, where we celebrate the joy, wonder, and thrill of mathematics.

In *Eureka Math* classrooms, learning is activated through rich experiences and dialogue. That new knowledge is best retained when it is reinforced with intentional practice. The *Learn, Practice, Succeed* book puts in students' hands the problem sets and fluency exercises they need to express and consolidate their classroom learning and master grade-level mathematics. Once students learn and practice, they know they can succeed.

What is in the Learn, Practice, Succeed *book?*

Fluency Practice: Our printed fluency activities utilize the format we call a Sprint. Instead of rote recall, Sprints use patterns across a sequence of problems to engage students in reasoning and to reinforce number sense while building speed and accuracy. Sprints are inherently differentiated, with problems building from simple to complex. The tempo of the Sprint provides a low-stakes adrenaline boost that increases memory and automaticity.

Classwork: A carefully sequenced set of examples, exercises, and reflection questions support students' in-class experiences and dialogue. Having classwork preprinted makes efficient use of class time and provides a written record that students can refer to later.

Exit Tickets: Students show teachers what they know through their work on the daily Exit Ticket. This check for understanding provides teachers with valuable real-time evidence of the efficacy of that day's instruction, giving critical insight into where to focus next.

Homework Helpers and Problem Sets: The daily Problem Set gives students additional and varied practice and can be used as differentiated practice or homework. A set of worked examples, Homework Helpers, support students' work on the Problem Set by illustrating the modeling and reasoning the curriculum uses to build understanding of the concepts the lesson addresses.

Homework Helpers and Problem Sets from prior grades or modules can be leveraged to build foundational skills. When coupled with *Affirm®*, *Eureka Math*'s digital assessment system, these Problem Sets enable educators to give targeted practice and to assess student progress. Alignment with the mathematical models and language used across *Eureka Math* ensures that students notice the connections and relevance to their daily instruction, whether they are working on foundational skills or getting extra practice on the current topic.

Where can I learn more about Eureka Math *resources?*

The Great Minds® team is committed to supporting students, families, and educators with an ever-growing library of resources, available at eureka-math.org. The website also offers inspiring stories of success in the *Eureka Math* community. Share your insights and accomplishments with fellow users by becoming a *Eureka Math* Champion.

Best wishes for a year filled with "aha" moments!

Jill Diniz

Jill Diniz
Chief Academic Officer, Mathematics
Great Minds

Contents

Module 2: Arithmetic Operations Including Division of Fractions

$$\frac{4}{1} \quad \frac{1}{4}$$ Reciprocals of each other

$$\frac{3}{1} \quad \frac{1}{3}$$

$$\frac{6}{1} \quad \frac{1}{6}$$

Notice: Dividing by 4 is same as Multiplying by $\frac{1}{4}$!

Opening Exercise

$16 \div 4$
④

$\frac{4}{1} \frac{16}{1} \times \frac{1}{4} \frac{1}{1} = \frac{4}{1}$ same as   ④

"of" means ×

reciprocal of 4

A

Write a division sentence to solve each problem.

1. 8 gallons of batter are poured equally into 4 bowls. How many gallons of batter are in each bowl?

2. 1 gallon of batter is poured equally into 4 bowls. How many gallons of batter are in each bowl?

Write a division sentence *and* draw a model to solve.

3. 3 gallons of batter are poured equally into 4 bowls. How many gallons of batter are in each bowl?

① 8 ÷ 4 = 2 gal.

② 1 ÷ 4 = $\frac{1}{4}$ gal.

③ 3 ÷ 4 = $\frac{3}{4}$ gal.

$\frac{1}{4} + \frac{1}{4} + \frac{1}{4} = \frac{3}{4}$

$\frac{3}{4}$ in each bowl

$\begin{array}{r} \times\ .75 \\ 4\ \overline{)3.00} \\ \underline{2\ 8} \downarrow \\ 20 \\ \underline{-20} \\ 0 \end{array}$

B

Write a multiplication sentence to solve each problem.

1. One fourth of an 8-gallon pail is poured out. How many gallons are poured out?

2. One fourth of a 1-gallon pail is poured out. How many gallons are poured out?

Write a multiplication sentence *and* draw a model to solve.

3. One fourth of a 3-gallon pail is poured out. How many gallons are poured out?

① $\frac{1}{4}$ × 8 = 2 gal.

② $\frac{1}{4}$ × 1 = $\frac{1}{4}$ gal.

③ $\frac{1}{4}$ × 3 = $\frac{3}{4}$ gal.

$\frac{1}{4} + \frac{1}{4} + \frac{1}{4} = \frac{3}{4}$

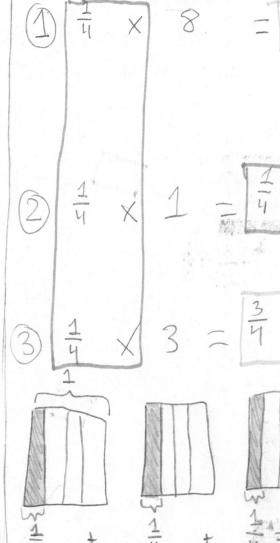

Example 1

$\frac{3}{4}$ gallon of batter is poured equally into 2 bowls. How many gallons of batter are in each bowl?

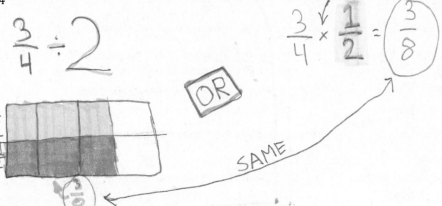

Example 2

$\frac{3}{4}$ pan of lasagna is shared equally by 6 friends. What fraction of the pan will each friend get?

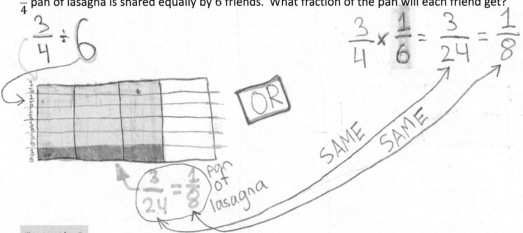

Example 3

A rope of length $\frac{2}{5}$ m is cut into 4 equal cords. What is the length of each cord?

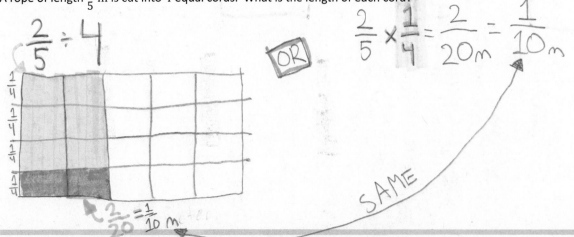

Lesson 1:   Interpreting Division of a Fraction by a Whole Number–Visual Models

EUREKA MATH®

Exercises 1–6

Fill in the blanks to complete the equation. Then, find the quotient and draw a model to support your solution.

1. $\frac{1}{2} \div 3 = \frac{\square}{3} \times \frac{1}{2}$ ① $\frac{1}{2} \div 3$

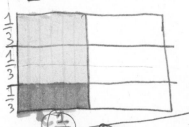

$\frac{1}{2} \div 3$

$\frac{1}{2} \times \frac{1}{3} = \boxed{\frac{1}{6}}$

OR

SAME

2. $\frac{1}{3} \div 4 = \frac{1}{4} \times \frac{\square}{}$ ② $\frac{1}{3} \div 4$

OR $\frac{1}{3} \times \frac{1}{4} = \boxed{\frac{1}{12}}$

SAME

$\frac{1}{12}$

Find the value of each of the following. Rewrite as Mult. & Solve

3. $\frac{1}{4} \div 5$

$\frac{1}{4} \times \frac{1}{5} = \boxed{\frac{1}{20}}$

4. $\frac{3}{4} \div 5$

$\frac{3}{4} \times \frac{1}{5} = \boxed{\frac{3}{20}}$

5. $\frac{1}{5} \div 4$

$\frac{1}{5} \times \frac{1}{4} = \boxed{\frac{1}{20}}$

Solve. Draw a model to support your solution. Rewrite as Mult.

6. $\frac{3}{5}$ pt. of juice is poured equally into 6 glasses. How much juice is in each glass?

$\frac{3}{5} \div 6$

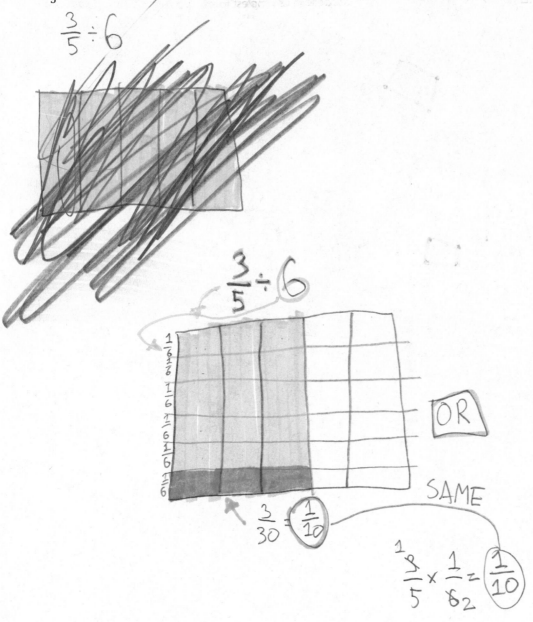

$\frac{3}{5} \div 6$

OR

$\frac{3}{30} = \frac{1}{10}$

SAME

$\frac{3}{5} \times \frac{1}{6} = \frac{1}{10}$

EUREKA MATH

Name _____   Date _____

Write an equivalent multiplication expression. Then, find the quotient in its simplest form. Use a model to support your response.

1. $\frac{1}{4} \div 2$

2. $\frac{2}{3} \div 6$

Find the value of each of the following in its simplest form.

1. $\frac{1}{2} \div 4$

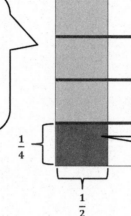

To divide by four, I can create four rows. From the model, I can see that I am finding $\frac{1}{4}$ of $\frac{1}{2}$. I see that $\frac{1}{2} \div 4$ is the same as $\frac{1}{2} \times \frac{1}{4}$.

The diagram begins with one whole unit. I can divide it into two equal parts (columns) and shade one part to represent $\frac{1}{2}$.

The shared area (dark blue) is one out of eight total pieces, or $\frac{1}{8}$.

$\frac{1}{2} \div 4 = \frac{1}{2} \times \frac{1}{4} = \frac{1}{8}$

2. Three loads of sand weigh $\frac{3}{4}$ tons. Find the weight of 1 load of sand.

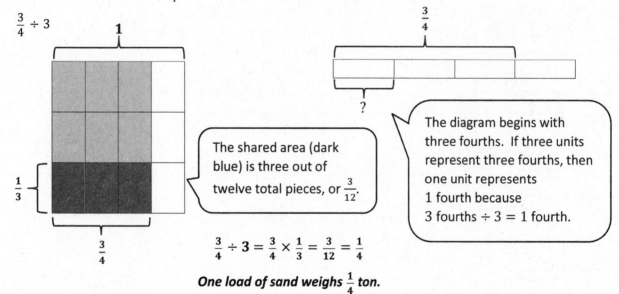

$\frac{3}{4} \div 3$

The shared area (dark blue) is three out of twelve total pieces, or $\frac{3}{12}$.

The diagram begins with three fourths. If three units represent three fourths, then one unit represents 1 fourth because 3 fourths ÷ 3 = 1 fourth.

$\frac{3}{4} \div 3 = \frac{3}{4} \times \frac{1}{3} = \frac{3}{12} = \frac{1}{4}$

One load of sand weighs $\frac{1}{4}$ ton.

3. Sammy cooked $\frac{1}{6}$ the amount of chicken he bought. He plans on cooking the rest equally over the next four days.

   a. What fraction of the chicken will Sammy cook each day?

$$\frac{6}{6} - \frac{1}{6} = \frac{5}{6}$$

> I begin with the whole amount of chicken, $\frac{6}{6}$, and then take away the $\frac{1}{6}$ he cooked.

> I divide the remaining $\frac{5}{6}$ by 4 to find the fraction for each day.

$$\frac{5}{6} \div 4 = \frac{5}{6} \times \frac{1}{4} = \frac{5}{24}$$

   Each day Sammy will cook $\frac{5}{24}$ the amount of chicken he bought.

   b. If Sammy has 48 pieces of chicken, how many pieces will he cook on Wednesday and Thursday?

   $\frac{5}{24}(48) = 10$; *he will cook 10 pieces each day, so $10 + 10 = 20$. He will cook 20 pieces of chicken on wednesday and Thursday.*

4. Sandra cooked $\frac{1}{3}$ of her sausages and put $\frac{1}{4}$ of the remaining sausages in the refrigerator to cook later. The rest of the sausages she divided equally into 2 portions and placed in the freezer.

   a. What fraction of sausage was in each container that went in the freezer?

$$\frac{3}{3} - \frac{1}{3} = \frac{2}{3}$$

> $\frac{1}{3}$ is cooked, so there are $\frac{2}{3}$ remaining.

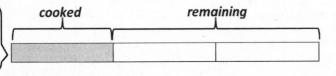

cooked   remaining

> To find a fourth of the remaining, I need to divide the remaining $\frac{2}{3}$ into 4 equal pieces.

$$\frac{2}{3} \div 4 = \frac{2}{3} \times \frac{1}{4} = \frac{2}{12} = \frac{1}{6}$$

> The darkest shaded value is $\frac{1}{4}$ the amount of the tape diagram.

> To find half of the remaining $\frac{6}{12}$, I need to divide by two.

$$\frac{6}{12} \div 2 = \frac{6}{12} \times \frac{1}{2} = \frac{6}{24} = \frac{3}{12} = \frac{1}{14}$$

   Each container that went in the freezer has $\frac{1}{4}$ of Sandra's sausages.

EUREKA MATH

b. If Sandra placed 20 sausages in the freezer, how many sausages did she start with?

$20 \div \frac{6}{12}$, *or* $20 \div \frac{1}{2}$

20 is $\frac{1}{2}$ of what size?

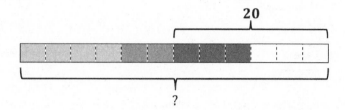

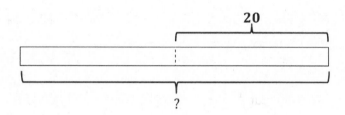

1 unit = 20

2 units = 2 × 20 = 40

Sandra started with 40 sausages.

Find the value of each of the following in its simplest form. *Rewrite as Mult. and Solve*

1.

a. $\frac{1}{3} \div 4$

$\frac{1}{3} \times \frac{1}{4} = \boxed{\frac{1}{12}}$

b. $\frac{2}{5} \div 4$

$\frac{2}{5} \times \frac{1}{4} = \boxed{\frac{1}{10}}$

c. $\frac{4}{7} \div 4$

$\frac{4}{7} \times \frac{1}{4} = \boxed{\frac{1}{7}}$

2.

a. $\frac{2}{5} \div 3$

$\frac{2}{5} \times \frac{1}{3} = \boxed{\frac{2}{15}}$

b. $\frac{5}{6} \div 5$

$\frac{5}{6} \times \frac{1}{5} = \boxed{\frac{1}{6}}$

c. $\frac{5}{8} \div 10$

$\frac{5}{8} \times \frac{1}{10} = \boxed{\frac{1}{16}}$

3.

a. $\frac{6}{7} \div 3$

$\frac{6}{7} \times \frac{1}{3} = \boxed{\frac{2}{7}}$

b. $\frac{10}{8} \div 5$

$\frac{10}{8} \times \frac{1}{5} = \frac{2}{8} = \boxed{\frac{1}{4}}$

c. $\frac{20}{6} \div 2$

$\frac{20}{6} \times \frac{1}{2} = \frac{10}{6} = 1\frac{4}{6} = \boxed{1\frac{2}{3}}$

4. 4 loads of stone weigh $\frac{2}{3}$ ton. Find the weight of 1 load of stone.

5. What is the width of a rectangle with an area of $\frac{5}{8}$ in² and a length of 10 inches?

6. Lenox ironed $\frac{1}{4}$ of the shirts over the weekend. She plans to split the remainder of the work equally over the next 5 evenings.

a. What fraction of the shirts will Lenox iron each day after school?

b. If Lenox has 40 shirts, how many shirts will she need to iron on Thursday and Friday?

7. Bo paid bills with $\frac{1}{2}$ of his paycheck and put $\frac{1}{5}$ of the remainder in savings. The rest of his paycheck he divided equally among the college accounts of his 3 children.

a. What fraction of his paycheck went into each child's account?

b. If Bo deposited $400 in each child's account, how much money was in Bo's original paycheck?

Example 1

Question # _____

Write it as a division expression.   _____

Write it as a multiplication expression.   _____

Make a rough draft of a model to represent the problem:

As you travel to each model, be sure to answer the following questions:

Original Question	Corresponding Division Expression	Corresponding Multiplication Expression	Write an Equation Showing the Equivalence of the Two Expressions.
1. How many $\frac{1}{2}$ miles are in 12 miles?			
2. How many quarter hours are in 5 hours?			
3. How many $\frac{1}{3}$ cups are in 9 cups?			
4. How many $\frac{1}{8}$ pizzas are in 4 pizzas?			
5. How many one-fifths are in 7 wholes?			

Lesson 2: Interpreting Division of a Whole Number by a Fraction—Visual Models

EUREKA MATH®

Example 2

Molly has 9 cups of flour. If this is $\frac{3}{4}$ of the number she needs to make bread, how many cups does she need?

    a.   Construct the tape diagram by reading it backward. Draw a tape diagram and label the unknown.

    b.   Next, shade in $\frac{3}{4}$.

    c.   Label the shaded region to show that 9 is equal to $\frac{3}{4}$ of the total.

    d.   Analyze the model to determine the quotient.

Exercises 1–5

1. A construction company is setting up signs on 2 miles of road. If the company places a sign every $\frac{1}{4}$ mile, how many signs will it use?

2. George bought 4 submarine sandwiches for a birthday party. If each person will eat $\frac{2}{3}$ of a sandwich, how many people can George feed?

3. Miranda buys 6 pounds of nuts. If she puts $\frac{3}{4}$ pound in each bag, how many bags can she make?

Lesson 2: Interpreting Division of a Whole Number by a
Fraction–Visual Models

EUREKA
MATH

4. Margo freezes 8 cups of strawberries. If this is $\frac{2}{3}$ of the total strawberries that she picked, how many cups of strawberries did Margo pick?

5. Regina is chopping up wood. She has chopped 10 logs so far. If the 10 logs represent $\frac{5}{8}$ of all the logs that need to be chopped, how many logs need to be chopped in all?

Name _____   Date _____

Solve each division problem using a model.

1. Henry bought 4 pies, which he plans to share with a group of his friends. If there is exactly enough to give each member of the group one-sixth of a pie, how many people are in the group?

2. Rachel finished $\frac{3}{4}$ of the race in 6 hours. How long was the entire race?

1. Ken used $\frac{5}{6}$ of his wrapping paper to wrap gifts. If he used 15 feet of wrapping paper, how much did he start with?

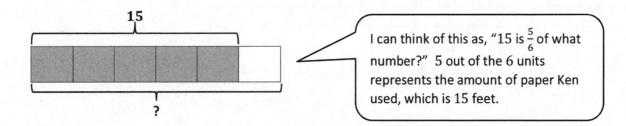

I can think of this as, "15 is $\frac{5}{6}$ of what number?" 5 out of the 6 units represents the amount of paper Ken used, which is 15 feet.

$15 \div \frac{5}{6}$

5 *units* = 15

1 *unit* = 15 ÷ 5 = 3

6 *units* = 6 × 3 = 18

I can divide 15 by 5 to determine the value of one unit. I need to find the value of one unit to determine the value of all six units.

Ken started with 18 feet of wrapping paper.

2. Robbie has 4 meters of ribbon. He cuts the ribbon into pieces $\frac{1}{3}$ meters long. How many pieces will he make?

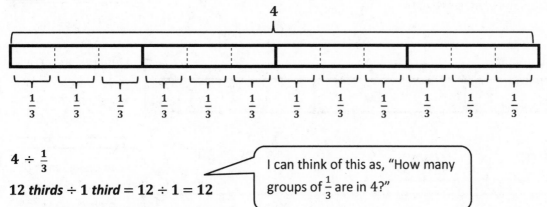

$4 \div \frac{1}{3}$

12 *thirds* ÷ 1 *third* = 12 ÷ 1 = 12

Robbie will make 12 pieces of ribbon.

I can think of this as, "How many groups of $\frac{1}{3}$ are in 4?"

3. Savannah spent $\frac{4}{5}$ of her money on clothes before spending $\frac{1}{3}$ of the remaining money on accessories. If the accessories cost $15, how much money did she have to begin with?

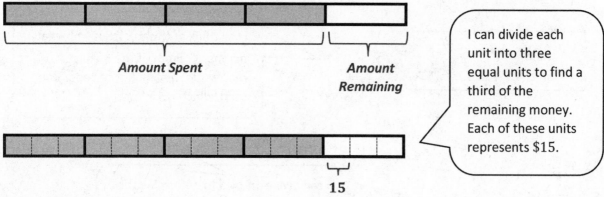

Amount Spent   **Amount Remaining**

15

> I can divide each unit into three equal units to find a third of the remaining money. Each of these units represents $15.

1 _unit_ = 15

15 _units_ = 15 × 15 = 225

Savannah had $225 _at first._

4. Isa's class was surveyed about their favorite foods. $\frac{1}{3}$ of the students preferred pizza, $\frac{1}{6}$ of the students preferred hamburgers, and $\frac{1}{2}$ of the remaining students preferred tacos. If 9 students preferred tacos, how many students were surveyed?

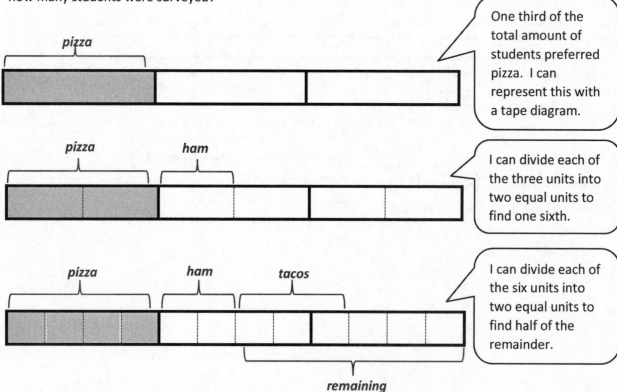

pizza

> One third of the total amount of students preferred pizza. I can represent this with a tape diagram.

pizza   *ham*

> I can divide each of the three units into two equal units to find one sixth.

pizza   *ham*   *tacos*

> I can divide each of the six units into two equal units to find half of the remainder.

remaining

Lesson 2:   Interpreting Division of a Whole Number by a
                                   Fraction–Visual Models

                        © 2019 Great Minds®. eureka-math.org

EUREKA MATH

3 *units* = 9

1 *unit* = 9 ÷ 3 = 3

12 *units* = 12 × 3 = 36

There were 36 students surveyed.

5. Caroline received her pay for the week. She spent $\frac{1}{4}$ of her pay on bills and deposited the remainder of the money equally into 2 bank accounts.

   a. What fraction of her pay did each bank account receive?

$$1 - \frac{1}{4} = \frac{3}{4}$$

$$\frac{3}{4} \div 2 = \frac{3}{4} \times \frac{1}{2} = \frac{3}{8}$$

> I need to start with the total amount of her pay, which I can represent with 1 whole.

   Each bank account received $\frac{3}{8}$ of her pay.

   b. If Caroline deposited $60 into each bank account, how much did she receive in her pay?

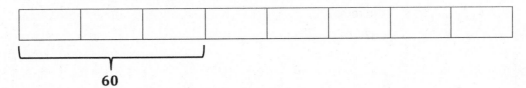

   60

   3 *units* = 60

   1 *unit* = 60 ÷ 3 = 20

   8 *units* = 8 × 20 = 160

   Caroline received $160 in her pay.

Rewrite each problem as a multiplication question. Model your answer.

1. Nicole used $\frac{3}{8}$ of her ribbon to wrap a present. If she used 6 feet of ribbon for the present, how much ribbon did Nicole have at first?

2. A Boy Scout has 3 meters of rope. He cuts the rope into cords $\frac{3}{5}$ m long. How many cords will he make?

3. 12 gallons of water fill a tank to $\frac{3}{4}$ capacity.

    a. What is the capacity of the tank?
    b. If the tank is then filled to capacity, how many half-gallon bottles can be filled with the water in the tank?

4. Hunter spent $\frac{2}{3}$ of his money on a video game before spending half of his remaining money on lunch. If his lunch costs $10, how much money did he have at first?

5. Students were surveyed about their favorite colors. $\frac{1}{4}$ of the students preferred red, $\frac{1}{8}$ of the students preferred blue, and $\frac{3}{5}$ of the remaining students preferred green. If 15 students preferred green, how many students were surveyed?

6. Mr. Scruggs got some money for his birthday. He spent $\frac{1}{5}$ of it on dog treats. Then, he divided the remainder equally among his 3 favorite charities.

    a. What fraction of his money did each charity receive?
    b. If he donated $60 to each charity, how much money did he receive for his birthday?

Opening Exercise

Draw a model to represent 12 ÷ 3.

Create a question or word problem that matches your model.

$$\frac{8}{9} \div \frac{2}{9}$$

Write the expression in unit form, and then draw a model to solve.

Example 2

$$\frac{9}{12} \div \frac{3}{12}$$

Write the expression in unit form, and then draw a model to solve.

Example 3

$$\frac{7}{9} \div \frac{3}{9}$$

Write the expression in unit form, and then draw a model to solve.

Lesson 3:   Interpreting and Computing Division of a Fraction by
a Fraction—More Models

EUREKA
MATH

Exercises 1–6

Write an expression to represent each problem. Then, draw a model to solve.

1. How many fourths are in 3 fourths?

2. $\dfrac{4}{5} \div \dfrac{2}{5}$

3. $\dfrac{9}{4} \div \dfrac{3}{4}$

4. $\dfrac{7}{8} \div \dfrac{2}{8}$

5. $\dfrac{13}{10} \div \dfrac{2}{10}$

6. $\dfrac{11}{9} \div \dfrac{3}{9}$

Lesson 3:   Interpreting and Computing Division of a Fraction by
                      a Fraction—More Models

EUREKA
MATH

Lesson Summary

When dividing a fraction by a fraction with the same denominator, we can use the general rule $\dfrac{a}{b} \div \dfrac{b}{c} = \dfrac{a}{b}$.

Name _____   Date _____

Find the quotient. Draw a model to support your solution.

1. $\dfrac{9}{4} \div \dfrac{3}{4}$

2. $\dfrac{7}{3} \div \dfrac{2}{3}$

Rewrite the expression in unit form. Find the quotient. Draw a model to support your answer.

1. $\frac{6}{8} \div \frac{2}{8}$

6 eighths ÷ 2 eighths = 6 ÷ 2 = 3

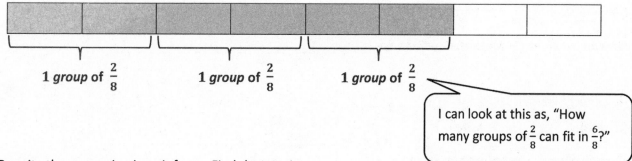

1 group of $\frac{2}{8}$ 1 group of $\frac{2}{8}$ 1 group of $\frac{2}{8}$

I can look at this as, "How many groups of $\frac{2}{8}$ can fit in $\frac{6}{8}$?"

Rewrite the expression in unit form. Find the quotient.

2. $\frac{7}{6} \div \frac{4}{6}$

7 sixths ÷ 4 sixths = 7 ÷ 4 = $\frac{7}{4}$ = $1\frac{3}{4}$

The units are the same in the dividend and divisor. I can easily divide the numerators.

Represent the division expression in unit form. Find the quotient.

3. A biker is $\frac{6}{7}$ miles from the finish line. If he can travel $\frac{5}{7}$ miles in one minute, how long until he reaches the finish line?

$\frac{6}{7} \div \frac{5}{7}$ = **6 sevenths ÷ 5 sevenths** = $6 \div 5$ = $\frac{6}{5}$ = $1\frac{1}{5}$

It will take him $1\frac{1}{5}$ minutes, or 1 minute and 12 seconds, to reach the finish line.

A seamstress has 5.2 feet of ribbon.

> Since this is a mixed number, she can only cut 8 whole strips.

a. How many $\frac{6}{10}$ feet strips of ribbon can she cut?

$5.2 = 52$ *tenths;* $\frac{6}{10} = 6$ *tenths;* 52 *tenths* $\div 6$ *tenths* $= 52 \div 6 = 8\frac{4}{6} = 8\frac{2}{3}$

She can cut eight $\frac{6}{10}$ *feet strips of ribbon.*

> I can determine eight strips of $\frac{6}{10}$ feet of ribbon by multiplying $\frac{6}{10}$ by 8.
> 6 tenths $\times$ 8 = 48 tenths

b. How much ribbon is left over?

52 *tenths* $- 48$ *tenths* $= 4$ *tenths*

She will have $\frac{4}{10}$ *feet of ribbon left over.*

For the following exercises, rewrite the division expression in unit form. Then, find the quotient. Draw a model to support your answer.

1. $\dfrac{4}{5} \div \dfrac{1}{5}$

2. $\dfrac{8}{9} \div \dfrac{4}{9}$

3. $\dfrac{15}{4} \div \dfrac{3}{4}$

4. $\dfrac{13}{5} \div \dfrac{4}{5}$

Rewrite the expression in unit form, and find the quotient.

5. $\dfrac{10}{3} \div \dfrac{2}{3}$

6. $\dfrac{8}{5} \div \dfrac{3}{5}$

7. $\dfrac{12}{7} \div \dfrac{12}{7}$

Represent the division expression using unit form. Find the quotient. Show all necessary work.

8. A runner is $\dfrac{7}{8}$ mile from the finish line. If she can travel $\dfrac{3}{8}$ mile per minute, how long will it take her to finish the race?

9. An electrician has 4.1 meters of wire.

   a. How many strips $\dfrac{7}{10}$ m long can he cut?

   b. How much wire will he have left over?

10. Saeed bought $21\dfrac{1}{2}$ lb. of ground beef. He used $\dfrac{1}{4}$ of the beef to make tacos and $\dfrac{2}{3}$ of the remainder to make quarter-pound burgers. How many burgers did he make?

11. A baker bought some flour. He used $\dfrac{2}{5}$ of the flour to make bread and used the rest to make batches of muffins.

    If he used 16 lb. of flour making bread and $\dfrac{2}{3}$ lb. for each batch of muffins, how many batches of muffins did he make?

Opening Exercise

Write at least three equivalent fractions for each fraction below.

a. $\dfrac{2}{3}$

b. $\dfrac{10}{12}$

Molly has $1\dfrac{3}{8}$ cups of strawberries. She needs $\dfrac{3}{8}$ cup of strawberries to make one batch of muffins. How many batches can Molly make?

Use a model to support your answer.

Example 2

Molly's friend, Xavier, also has $\frac{11}{8}$ cups of strawberries. He needs $\frac{3}{4}$ cup of strawberries to make a batch of tarts. How many batches can he make? Draw a model to support your solution.

Example 3

Find the quotient: $\frac{6}{8} \div \frac{2}{8}$. Use a model to show your answer.

Lesson 4:   Interpreting and Computing Division of a Fraction by a Fraction—More Models

EUREKA MATH

Example 4

Find the quotient: $\frac{3}{4} \div \frac{2}{3}$. Use a model to show your answer.

Exercises 1–5

Find each quotient.

1. $\frac{6}{2} \div \frac{3}{4}$

2. $\frac{2}{3} \div \frac{2}{5}$

3. $\frac{7}{8} \div \frac{1}{2}$

4. $\frac{3}{5} \div \frac{1}{4}$

Lesson 4:   Interpreting and Computing Division of a Fraction by
a Fraction–More Models

EUREKA MATH®

5. $\dfrac{5}{4} \div \dfrac{1}{3}$

Name _____   Date _____

Calculate each quotient. If needed, draw a model.

1. $\dfrac{9}{4} \div \dfrac{3}{8}$

2. $\dfrac{3}{5} \div \dfrac{2}{3}$

Calculate the quotient. If needed, draw a model.

1. $\frac{2}{5} \div \frac{2}{3}$

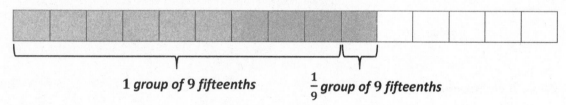

These fractions do not have the same denominator, or unit. I need to create like denominators to divide the numerators.

6 fifteenths ÷ 10 fifteenths = 6 ÷ 10 = $\frac{6}{10}$, or $\frac{3}{5}$

2. $\frac{2}{3} \div \frac{3}{5}$

10 fifteenths ÷ 9 fifteenths = 10 ÷ 9 = $\frac{10}{9}$ = $1\frac{1}{9}$

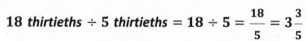

1 group of 9 fifteenths $\frac{1}{9}$ group of 9 fifteenths

3. $\frac{3}{5} \div \frac{1}{6}$

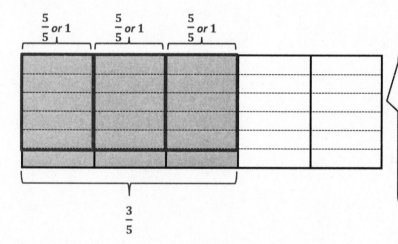

$\frac{5}{5}$ or 1 $\frac{5}{5}$ or 1 $\frac{5}{5}$ or 1

$\frac{3}{5}$

I can shade 3 out of 5 columns to represent $\frac{3}{5}$. To find how many groups of $\frac{1}{6}$ are in that amount, I can divide each column into 6 rows. There are 18 fifths. I can represent this as 3 wholes and 3 fifths, or $3\frac{3}{5}$.

18 thirtieths ÷ 5 thirtieths = 18 ÷ 5 = $\frac{18}{5}$ = $3\frac{3}{5}$

Lesson 4: Interpreting and Computing Division of a Fraction by a Fraction–More Models

4. $\dfrac{5}{6} \div \dfrac{1}{3}$

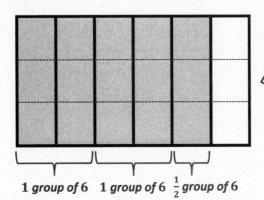

1 *group of* 6   1 *group of* 6 $\dfrac{1}{2}$ *group of* 6

I can shade 5 out of 6 columns to represent $\dfrac{5}{6}$. To find how many groups of $\dfrac{1}{3}$ are in that amount, I can divide each column into 3 rows. There are 15 sixths. I can represent this as 2 wholes and 3 sixths, or $2\dfrac{1}{2}$.

15 *eighteenths* $\div$ **6** *eighteenths* $= 15 \div 6 = \dfrac{15}{6} = 2\dfrac{1}{2}$

Lesson 4:   Interpreting and Computing Division of a Fraction by a Fraction—More Models

EUREKA MATH

Calculate the quotient. If needed, draw a model.

1. $\dfrac{8}{9} \div \dfrac{4}{9}$

2. $\dfrac{9}{10} \div \dfrac{4}{10}$

3. $\dfrac{3}{5} \div \dfrac{1}{3}$

4. $\dfrac{3}{4} \div \dfrac{1}{5}$

Opening Exercise

Tape Diagram:

$\dfrac{8}{9} \div \dfrac{2}{9}$

Number Line:

Molly's friend, Xavier, also has $\dfrac{11}{8}$ cups of strawberries. He needs $\dfrac{3}{4}$ cup of strawberries to make a batch of tarts. How many batches can he make? Draw a model to support your solution.

Example 1

$$\frac{1}{2} \div \frac{1}{8}$$

Step 1: Decide on an interpretation.

Step 2: Draw a model.

Step 3: Find the answer.

Step 4: Choose a unit.

Step 5: Set up a situation based upon the model.

Lesson 5:   Creating Division Stories

EUREKA MATH

Exercise 1

Using the same dividend and divisor, work with a partner to create your own story problem. You may use the same unit, but your situation must be unique. You could try another unit such as ounces, yards, or miles if you prefer.

Example 2

$$\frac{3}{4} \div \frac{1}{2}$$

Step 1: Decide on an interpretation.

Step 2: Draw a diagram.

Step 3: Find the answer.

Step 4: Choose a unit.

Step 5: Set up a situation based on the model.

Exercise 2

Using the same dividend and divisor, work with a partner to create your own story problem. You may use the same unit, but your situation must be unique. You could try another unit such as cups, yards, or miles if you prefer.

Lesson Summary

The method of creating division stories includes five steps:

Step 1: Decide on an interpretation (measurement or partitive). Today we used measurement division.

Step 2: Draw a model.

Step 3: Find the answer.

Step 4: Choose a unit.

Step 5: Set up a situation based on the model. This means writing a story problem that is interesting, realistic, and short. It may take several attempts before you find a story that works well with the given dividend and divisor.

Name _____   Date _____

Write a story problem using the measurement interpretation of division for the following: $\frac{3}{4} \div \frac{1}{8} = 6$.

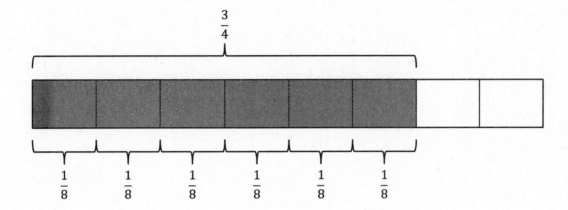

1. How many $\frac{1}{3}$ teaspoons of honey are in a recipe calling for $\frac{5}{6}$ teaspoons of honey?

$$\frac{5}{6} \div \frac{1}{3} = \frac{5}{6} \div \frac{2}{6}$$

5 sixths $\div$ 2 sixths $= 5 \div 2 = \frac{5}{2} = 2\frac{1}{2}$

There are $2\frac{1}{2}$ one-third teaspoons of honey in $\frac{5}{6}$ teaspoons.

2. Write a measurement story problem for $5 \div \frac{3}{5}$.

> I know that measurement interpretation means that I have to find out how many groups of $\frac{3}{5}$ are in 5.

How many $\frac{3}{5}$ cups of milk are in a recipe calling for 5 cups?

3. Fill in the blanks to complete the equation. Then, find the quotient, and draw a model to support your solution.

$$\frac{1}{3} \div 7 = \frac{1}{\Box} \text{ of } \frac{1}{3}$$

$$\frac{1}{3} \div 7 = \frac{1}{7} \text{ of } \frac{1}{3}$$

$$= \frac{1}{21}$$

> When I divide by 7, I know that is the same as taking a seventh, or multiplying by $\frac{1}{7}$. The word "of" tells me to multiply in this case.

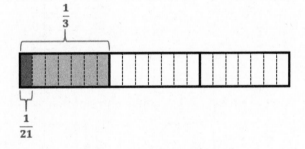

EUREKA
MATH®

4. Pam used 8 loads of soil to cover $\frac{4}{5}$ of her garden. How many loads of soil will she need to cover the entire garden?

   4 *units* = 8

   1 *unit* = 8 ÷ 4 = 2

   5 *units* = 5 × 2 = 10

   > I can use the partitive interpretation of division here since I know both parts and need to determine the total amount.

   Pam needs 10 loads of soil to cover the entire garden.

5. Becky plans to run 3 miles on the track. Each lap is $\frac{1}{4}$ miles. How many laps will Becky run?

   $3 \div \frac{1}{4} = 12$ ***fourths* ÷ 1 *fourth* = $12 \div 1 = \frac{12}{1} = 12$. *Becky will run 12 laps*.**

6. Kaliah spent $\frac{2}{3}$ of her money on an outfit. She spent $\frac{3}{8}$ of the remaining money on a necklace. If she has $15 left, how much did the outfit cost?

   $$\frac{3}{3} - \frac{2}{3} = \frac{1}{3}$$

   $$\frac{1}{3} \times \frac{3}{8} = \frac{1}{8}$$

   $$\frac{2}{3} + \frac{1}{8} = \frac{16}{24} + \frac{3}{24} = \frac{19}{24}$$

   $$\frac{24}{24} - \frac{19}{24} = \frac{5}{24}$$

   15 *is* $\frac{5}{24}$ *of what number*?

   5 *units* = 15

   1 *unit* = 15 ÷ 5 = 3

   16 *units* = 16 × 3 = 48

   ***The outfit cost* $48.**

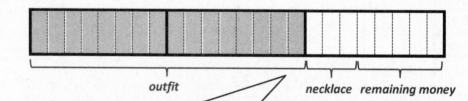

   outfit   necklace remaining money
                                   $15

   > $\frac{2}{3}$ is shaded in my diagram. What is left over is $\frac{1}{3}$. Three eighths of that is spent on the necklace. The leftover is $\frac{5}{8}$. If I split the remaining third into eight equal parts, I need to split each of the other two thirds into eight equal parts. The entire amount is now in 24 parts.

   > 5 units out of 24 represents the $15 left over. I can use unit form to determine what one unit represents.

EUREKA MATH

Solve.

1. How many sixteenths are in $\frac{15}{16}$?

2. How many $\frac{1}{4}$ teaspoon doses are in $\frac{7}{8}$ teaspoon of medicine?

3. How many $\frac{2}{3}$ cups servings are in a 4 cup container of food?

4. Write a measurement division story problem for $6 \div \frac{3}{4}$.

5. Write a measurement division story problem for $\frac{5}{12} \div \frac{1}{6}$.

6. Fill in the blank to complete the equation. Then, find the quotient and draw a model to support your solution.

    a. $\frac{1}{2} \div 5 = \frac{1}{\Box}$ of $\frac{1}{2}$ b. $\frac{3}{4} \div 6 = \frac{1}{\Box}$ of $\frac{3}{4}$

7. $\frac{4}{5}$ of the money collected from a fundraiser was divided equally among 8 grades. What fraction of the money did each grade receive?

8. Meyer used 6 loads of gravel to cover $\frac{2}{5}$ of his driveway. How many loads of gravel will he need to cover his entire driveway?

9. An athlete plans to run 3 miles. Each lap around the school yard is $\frac{3}{7}$ mile. How many laps will the athlete run?

10. Parks spent $\frac{1}{3}$ of his money on a sweater. He spent $\frac{3}{5}$ of the remainder on a pair of jeans. If he has $36 left, how much did the sweater cost?

Example 1

Divide $50 \div \frac{2}{3}$.

Step 1: Decide on an interpretation.

Step 2: Draw a model.

Step 3: Find the answer.

Step 4: Choose a unit.

Step 5: Set up a situation based upon the model.

Exercise 1

Using the same dividend and divisor, work with a partner to create your own story problem. You may use the same unit, dollars, but your situation must be unique. You could try another unit, such as miles, if you prefer.

Example 2

Divide $\frac{1}{2} \div \frac{3}{4}$.

Step 1: Decide on an interpretation.

Step 2: Draw a model.

Lesson 6:   More Division Stories

EUREKA
MATH

Step 3: Find the answer.

Step 4: Choose a unit.

Step 5: Set up a situation based upon the model.

Exercise 2

Using the same dividend and divisor, work with a partner to create your own story problem. Try a different unit.

Name _____ Date _____

Write a story problem using the partitive interpretation of division for the following: $25 \div \frac{5}{8} = 40$.

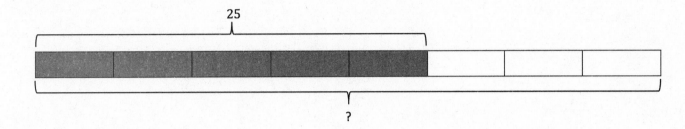

1. $\frac{5}{6}$ teaspoons is $\frac{1}{3}$ group of what size?

$$\frac{5}{6} \div \frac{1}{3}$$

5 sixths ÷ 2 sixths = $\frac{5}{2}$ = $2\frac{1}{2}$

$\frac{5}{6}$ **teaspoons is** $\frac{1}{3}$ **group of** $2\frac{1}{2}$ **teaspoons.**

> In partitive division, I know the parts and need to find the total amount. I can choose the unit of feet and create a story.

2. Write a partitive division story problem for $\frac{7}{10} \div \frac{1}{5}$.

Brendan had $\frac{7}{10}$ foot of rope. This is $\frac{1}{5}$ the amount he needs. How much rope does he need in all?

3. Fill in the blanks to complete the equation. Then, find the quotient, and draw a model to support your solution.

$$\frac{5}{6} \div 4 = \frac{\square}{4} \text{ of } \frac{5}{6}$$
$$\frac{5}{6} \div 4 = \frac{1}{4} \text{ of } \frac{5}{6}$$

> I can think of this as what is $\frac{1}{4}$ of $\frac{5}{6}$? $\frac{5}{6}$ is the total. I am looking for the part.

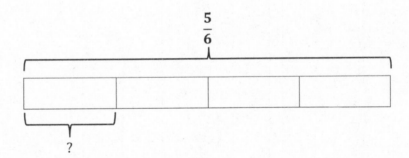

$\frac{5}{6}$

?

4 units = $\frac{5}{6}$

1 unit = $\frac{5}{6} \div 4 = \frac{5}{6} \times \frac{1}{4} = \frac{5}{24}$

4. Karrie cleaned $\frac{1}{5}$ of her house in 45 minutes. How long will it take her to clean the entire house?

$$45 \text{ min} \times \frac{1}{60} \frac{\text{hr}}{\text{min}} = \frac{45}{60} \text{ hr} = \frac{3}{4} \text{ hr}$$

$$\frac{3}{4} \div \frac{1}{5} = 15 \text{ twentieths} \div \text{twentieths} = \frac{15}{4} = 3\frac{3}{4}$$

> I can use conversions to determine the fraction of an hour that is represented by 45 minutes.

It will take Karrie $3\frac{3}{4}$ hours to clean the entire house.

> I can look at this as partitive division. I know it takes $\frac{3}{4}$ hours to clean $\frac{1}{5}$ of the house. I'm looking to find the total amount of hours needed to clean the whole house.

EUREKA MATH

Solve.

1. $\frac{15}{16}$ is 1 sixteenth groups of what size?

2. $\frac{7}{8}$ teaspoons is $\frac{1}{4}$ groups of what size?

3. A 4-cup container of food is $\frac{2}{3}$ groups of what size?

4. Write a partitive division story problem for $6 \div \frac{3}{4}$.

5. Write a partitive division story problem for $\frac{5}{12} \div \frac{1}{6}$.

6. Fill in the blank to complete the equation. Then, find the quotient, and draw a model to support your solution.

    a. $\frac{1}{4} \div 7 = \frac{1}{\square} \text{ of } \frac{1}{4}$

    b. $\frac{5}{6} \div 4 = \frac{1}{\square} \text{ of } \frac{5}{6}$

7. There is $\frac{3}{5}$ of a pie left. If 4 friends wanted to share the pie equally, how much would each friend receive?

8. In two hours, Holden completed $\frac{3}{4}$ of his race. How long will it take Holden to complete the entire race?

9. Sam cleaned $\frac{1}{3}$ of his house in 50 minutes. How many hours will it take him to clean his entire house?

10. It took Mario 10 months to beat $\frac{5}{8}$ of the levels on his new video game. How many years will it take for Mario to beat all the levels?

11. A recipe calls for $1\frac{1}{2}$ cups of sugar. Marley only has measuring cups that measure $\frac{1}{4}$ cup. How many times will Marley have to fill the measuring cup?

Example 1

Model the following using a partitive interpretation.

$$\frac{3}{4} \div \frac{2}{5}$$

Shade 2 of the 5 sections $\left(\frac{2}{5}\right)$.

Label the part that is known $\left(\frac{3}{4}\right)$.

Make notes below on the math sentences needed to solve the problem.

Example 2

Model the following using a measurement interpretation.

$\frac{3}{5} \div \frac{1}{4}$

Example 3

$\frac{2}{3} \div \frac{3}{4}$

Show the number sentences below.

Lesson 7:   The Relationship Between Visual Fraction Models
and Equations

EUREKA
MATH®

Lesson Summary

Connecting models of fraction division to multiplication through the use of reciprocals helps in understanding the *invert and multiply* rule. That is, given two fractions $\frac{a}{b}$ and $\frac{c}{d}$, we have the following:

$$\frac{a}{b} \div \frac{c}{d} = \frac{a}{b} \times \frac{d}{c}.$$

1 whole unit

| $\frac{1}{2}$ | | $\frac{1}{2}$ | |

| $\frac{1}{3}$ | $\frac{1}{3}$ | $\frac{1}{3}$ |

| $\frac{1}{4}$ | $\frac{1}{4}$ | $\frac{1}{4}$ | $\frac{1}{4}$ |

| $\frac{1}{5}$ | $\frac{1}{5}$ | $\frac{1}{5}$ | $\frac{1}{5}$ | $\frac{1}{5}$ |

| $\frac{1}{6}$ | $\frac{1}{6}$ | $\frac{1}{6}$ | $\frac{1}{6}$ | $\frac{1}{6}$ | $\frac{1}{6}$ |

| $\frac{1}{8}$ | $\frac{1}{8}$ | $\frac{1}{8}$ | $\frac{1}{8}$ | $\frac{1}{8}$ | $\frac{1}{8}$ | $\frac{1}{8}$ | $\frac{1}{8}$ |

| $\frac{1}{9}$ | $\frac{1}{9}$ | $\frac{1}{9}$ | $\frac{1}{9}$ | $\frac{1}{9}$ | $\frac{1}{9}$ | $\frac{1}{9}$ | $\frac{1}{9}$ | $\frac{1}{9}$ |

| $\frac{1}{10}$ | $\frac{1}{10}$ | $\frac{1}{10}$ | $\frac{1}{10}$ | $\frac{1}{10}$ | $\frac{1}{10}$ | $\frac{1}{10}$ | $\frac{1}{10}$ | $\frac{1}{10}$ | $\frac{1}{10}$ |

| $\frac{1}{12}$ | $\frac{1}{12}$ | $\frac{1}{12}$ | $\frac{1}{12}$ | $\frac{1}{12}$ | $\frac{1}{12}$ | $\frac{1}{12}$ | $\frac{1}{12}$ | $\frac{1}{12}$ | $\frac{1}{12}$ | $\frac{1}{12}$ | $\frac{1}{12}$ |

Name _____   Date _____

1. Write the reciprocal of the following numbers.

Number	$\frac{7}{10}$	$\frac{1}{2}$	5
Reciprocal			

2. Rewrite this division expression as an equivalent multiplication expression: $\frac{5}{8} \div \frac{2}{3}$.

3. Solve Problem 2. Draw a model to support your solution.

Invert and multiply to divide.

1. $\frac{6}{7} \div \frac{2}{3}$

$$\frac{6}{7} \div \frac{2}{3} = \frac{6}{7} \times \frac{3}{2} = \frac{18}{14} = \frac{9}{7}$$

> I know that $\frac{6}{7}$ is $\frac{2}{3}$ of a number. Two units is represented by $\frac{6}{7}$, so one unit is half of $\frac{6}{7}$. $\frac{6}{7} \times \frac{1}{2} = \frac{6}{14}$. Three units is $3 \times \frac{6}{14} = \frac{18}{14}$. I multiplied $\frac{6}{7}$ by 3 and by $\frac{1}{2}$. I know this is the same as multiplying $\frac{6}{7}$ by $\frac{3}{2}$.

2. Cody used $\frac{3}{4}$ of the gas in his tank. If $\frac{5}{7}$ of the tank of gas is used, how much gas did Cody start with?

> I know that this problem is asking me to determine "$\frac{5}{7}$ is $\frac{3}{4}$ of what number?"

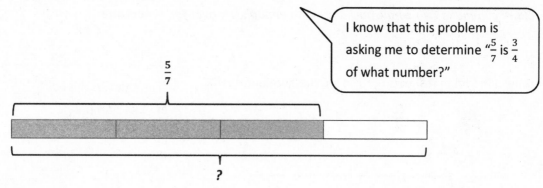

$\frac{5}{7}$ is $\frac{3}{4}$ of what number?

$\frac{5}{7} \div \frac{3}{4}$

3 units $= \frac{5}{7}$

1 unit $= \frac{5}{7} \div 3 = \frac{5}{7} \times \frac{1}{3} = \frac{5}{21}$

> This shows why I can invert and multiply the second factor.

4 units $= \frac{5}{21} \times 4 = \frac{20}{21}$

$\frac{5}{7}$ is $\frac{3}{4}$ of $\frac{20}{21}$.

Cody started with $\frac{20}{21}$ of a tank of gas.

3. Claire has 7 half-pound packages of trail mix. She wants to make packages that contain $1\frac{1}{2}$ pounds. How many packages can she make?

$1\frac{1}{2} = \frac{2}{2} + \frac{1}{2} = \frac{3}{2}$

> I need to represent this mixed number with a fraction and then invert and multiply.

$\frac{7}{2}$ **is how many** $\frac{3}{2}$**?**

$\frac{7}{2} \div \frac{3}{2} = \frac{7}{2} \times \frac{2}{3} = \frac{14}{6}$

$\frac{14}{6} = \frac{7}{3} = 2\frac{1}{3}$

Claire can make two whole packages with enough left over for $\frac{1}{3}$ package.

4. Draw a model that shows $\frac{3}{5} \div \frac{1}{2}$. Find the quotient.

> I can think of this as, "$\frac{3}{5}$ is $\frac{1}{2}$ of what number?"

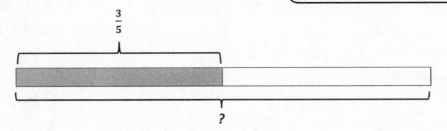

$$\frac{3}{5} \div \frac{1}{2} = \frac{3}{5} \times \frac{2}{1} = \frac{6}{5} = 1\frac{1}{5}$$

       Lesson 7:   The Relationship Between Visual Fraction Models and Equations

EUREKA MATH

Invert and multiply to divide.

1.

    a. $\frac{2}{3} \div \frac{1}{4}$ b. $\frac{2}{3} \div 4$ c. $4 \div \frac{2}{3}$

2.

    a. $\frac{1}{3} \div \frac{1}{4}$ b. $\frac{1}{8} \div \frac{3}{4}$ c. $\frac{9}{4} \div \frac{6}{5}$

3.

    a. $\frac{2}{3} \div \frac{3}{4}$ b. $\frac{3}{5} \div \frac{3}{2}$ c. $\frac{22}{4} \div \frac{2}{5}$

4. Summer used $\frac{2}{5}$ of her ground beef to make burgers. If she used $\frac{3}{4}$ pounds of beef, how much beef did she have at first?

5. Alistair has 5 half-pound chocolate bars. It takes $1\frac{1}{2}$ pounds of chocolate, broken into chunks, to make a batch of cookies. How many batches can Alistair make with the chocolate he has on hand?

6. Draw a model that shows $\frac{2}{5} \div \frac{1}{3}$. Find the answer as well.

7. Draw a model that shows $\frac{3}{4} \div \frac{1}{2}$. Find the answer as well.

Example 1: Introduction to Calculating the Quotient of a Mixed Number and a Fraction

a. Carli has $4\frac{1}{2}$ walls left to paint in order for all the bedrooms in her house to have the same color paint.

However, she has used almost all of her paint and only has $\frac{5}{6}$ of a gallon left.

How much paint can she use on each wall in order to have enough to paint the remaining walls?

b. Calculate the quotient.

$$\frac{2}{5} \div 3\frac{4}{7}$$

Exercise

Show your work for the memory game in the boxes provided below.

A.	
B.	
C.	
D.	
E.	
F.	
G.	
H.	
I.	
J.	
K.	
L.	

A.		B.	
$\dfrac{3}{4} \div 6\dfrac{2}{3}$	$\dfrac{9}{80}$	$\dfrac{1}{3} \div 4\dfrac{3}{4}$	$\dfrac{4}{57}$
C.		D.	
$\dfrac{2}{5} \div 1\dfrac{7}{8}$	$\dfrac{16}{75}$	$7\dfrac{1}{2} \div \dfrac{5}{6}$	9
E.		F.	
$3\dfrac{4}{7} \div \dfrac{5}{8}$	$5\dfrac{5}{7}$	$5\dfrac{5}{8} \div \dfrac{9}{10}$	$6\dfrac{1}{4}$
G.		H.	
$\dfrac{1}{4} \div 10\dfrac{11}{12}$	$\dfrac{3}{131}$	$5\dfrac{3}{4} \div \dfrac{5}{9}$	$10\dfrac{7}{20}$
I.		J.	
$3\dfrac{1}{5} \div \dfrac{2}{3}$	$4\dfrac{4}{5}$	$\dfrac{3}{5} \div 3\dfrac{1}{7}$	$\dfrac{21}{110}$
K.		L.	
$\dfrac{10}{13} \div 2\dfrac{4}{7}$	$\dfrac{35}{117}$	$2\dfrac{1}{4} \div \dfrac{7}{8}$	$2\dfrac{4}{7}$

Name _____ Date _____

Calculate the quotient.

1. $\dfrac{3}{4} \div 5\dfrac{1}{5}$

2. $\dfrac{3}{7} \div 2\dfrac{1}{2}$

3. $\dfrac{5}{8} \div 6\dfrac{5}{6}$

4. $\dfrac{5}{8} \div 8\dfrac{3}{10}$

Calculate each quotient.

1. $\dfrac{3}{7} \div 4\dfrac{1}{5}$

$$4\dfrac{1}{5} = \left(4 \times \dfrac{5}{5}\right) + \dfrac{1}{5}$$

$$\dfrac{20}{5} + \dfrac{1}{5} = \dfrac{21}{5}$$

$$\dfrac{3}{7} \div \dfrac{21}{5} = \dfrac{3}{7} \times \dfrac{5}{21} = \dfrac{15}{147} = \dfrac{5}{49}$$

> Before I divide, I need to change $4\dfrac{1}{5}$ into a fraction. I know that 4 can be represented as $\dfrac{20}{5}$. I can add that to $\dfrac{1}{5}$ to determine the equivalent fraction.

2. $5\dfrac{1}{3} \div \dfrac{5}{8}$

$$5\dfrac{1}{3} = \left(5 \times \dfrac{3}{3}\right) + \dfrac{1}{3}$$

$$\dfrac{15}{3} + \dfrac{1}{3} = \dfrac{16}{3}$$

$$\dfrac{16}{3} \div \dfrac{5}{8} = \dfrac{16}{3} \times \dfrac{8}{5} = \dfrac{128}{15} = 8\dfrac{8}{15}$$

> Before I divide, I need to change $5\dfrac{1}{3}$ into a fraction. I know that 5 can be represented as $\dfrac{15}{3}$. I can add that to $\dfrac{1}{3}$ to determine the equivalent fraction.

EUREKA MATH®

Lesson 8:   Dividing Fractions and Mixed Numbers

Calculate each quotient.

1. $\dfrac{2}{5} \div 3\dfrac{1}{10}$

2. $4\dfrac{1}{3} \div \dfrac{4}{7}$

3. $3\dfrac{1}{6} \div \dfrac{9}{10}$

4. $\dfrac{5}{8} \div 2\dfrac{7}{12}$

Example 1

$$25\frac{3}{10} + 376\frac{77}{100}$$

Example 2

$$426\frac{1}{5} - 275\frac{1}{2}$$

Exercises

Calculate each sum or difference.

1. Samantha and her friends are going on a road trip that is $245\frac{7}{50}$ miles long. They have already driven $128\frac{53}{100}$. How much farther do they have to drive?

2. Ben needs to replace two sides of his fence. One side is $367\frac{9}{100}$ meters long, and the other is $329\frac{3}{10}$ meters long. How much fence does Ben need to buy?

3. Mike wants to paint his new office with two different colors. If he needs $4\frac{4}{5}$ gallons of red paint and $3\frac{1}{10}$ gallons of brown paint, how much paint does he need in total?

EUREKA
MATH®

4. After Arianna completed some work, she figured she still had $78\frac{21}{100}$ pictures to paint. If she completed another $34\frac{23}{25}$ pictures, how many pictures does Arianna still have to paint?

Use a calculator to convert the fractions into decimals before calculating the sum or difference.

5. Rahzel wants to determine how much gasoline he and his wife use in a month. He calculated that he used $78\frac{1}{3}$ gallons of gas last month. Rahzel's wife used $41\frac{3}{8}$ gallons of gas last month. How much total gas did Rahzel and his wife use last month? Round your answer to the nearest hundredth.

Name _____   Date _____

Solve each problem. Show that the placement of the decimal is correct through either estimation or fraction calculation.

1. $382\frac{3}{10} - 191\frac{87}{100}$

2. $594\frac{7}{25} + 89\frac{37}{100}$

Find each sum or difference.

1. $426\frac{2}{10} - 215\frac{68}{100}$

   426.2 − 215.68

   426.20 − 215.68

It would be difficult to subtract the mixed numbers, so I can represent the numbers with decimals. From there, I can use the subtraction algorithm to find the difference.

$$
\begin{array}{r}
5 \quad\ \ 11\ 10 \\
4\ \ 2\ \ \cancel{6}\ .\ \cancel{2}\ \ \cancel{0} \\
-\quad 2\ \ 1\ \ 5\ .\ 6\ \ 8 \\
\hline
2\ \ 1\ \ 0\ .\ 5\ \ 2
\end{array}
$$

2. $627\frac{17}{25} + 18\frac{7}{10}$

   627.68 + 18.7

   627.68 + 18.70

It would be difficult to add the mixed numbers, so I can represent the numbers with decimals. From there, I can use the addition algorithm to find the sum.

$$
\begin{array}{r}
6\ \ 2\ \ 7\ .\ 6\ \ 8 \\
+\quad\ \ 1\ \ 8\ .\ 7\ \ 0 \\
\hline
6\ \ 4\ \ 6\ .\ 3\ \ 8
\end{array}
$$

1. Find each sum or difference.

   a. $381\frac{1}{10} - 214\frac{43}{100}$

   b. $32\frac{3}{4} - 12\frac{1}{2}$

   c. $517\frac{37}{50} + 312\frac{3}{100}$

   d. $632\frac{16}{25} + 32\frac{3}{10}$

   e. $421\frac{3}{50} - 212\frac{9}{10}$

2. Use a calculator to find each sum or difference. Round your answer to the nearest hundredth.

   a. $422\frac{3}{7} - 367\frac{5}{9}$

   b. $23\frac{1}{5} + 45\frac{7}{8}$

Opening Exercise

Calculate the product.

   a. 200×32.6

   b. 500×22.12

Use partial products and the distributive property to calculate the product.

200×32.6

Use partial products and the distributive property to calculate the area of the rectangular patio shown below.

22.12 ft.

500 ft.

Exercises

Use the boxes below to show your work for each station. Make sure that you are putting the solution for each station in the correct box.

Station One:

Station Two:

Station Three:

Station Four:

Station Five:

Lesson 10:   The Distributive Property and the Products of
Decimals

EUREKA
MATH®

Name _____   Date _____

Complete the problem using partial products.

500×12.7

Calculate the product using partial products.

1. 500×54.1

    $500(50) + 500(4) + 500(0.1)$

    $25,000 + 2,000 + 50$

    $27,050$

> I can decompose 54.1 into an addition expression. 54.1 is equal to the sum $50 + 4 + 0.1$. I can now distribute 500 to each addend in the expression: $500(50) + 500(4) + 500(0.1)$.

2. 13.5×200

    $200(10) + 200(3) + 200(0.5)$

    $2,000 + 600 + 100$

    $2,700$

> The commutative property allows me to switch the factors in the problem. 200×13.5
>
> I can decompose 13.5 into an addition expression. 13.5 is equal to the sum $10 + 3 + 0.5$. I can now distribute 200 to each addend in the expression: $200(10) + 200(3) + 200(0.5)$.

Calculate the product using partial products.

1. 400×45.2

2. 14.9×100

3. 200×38.4

4. 900×20.7

5. 76.2×200

Exploratory Challenge

You not only need to solve each problem, but your groups also need to prove to the class that the decimal in the product is located in the correct place. As a group, you are expected to present your informal proof to the class.

a.   Calculate the product. 34.62×12.8

b.   Xavier earns $11.50 per hour working at the nearby grocery store. Last week, Xavier worked for 13.5 hours. How much money did Xavier earn last week? Remember to round to the nearest penny.

Discussion

Record notes from the Discussion in the box below.

Exercises

1. Calculate the product. 324.56×54.82

2. Kevin spends \$11.25 on lunch every week during the school year. If there are 35.5 weeks during the school year, how much does Kevin spend on lunch over the entire school year? Remember to round to the nearest penny.

3. Gunnar's car gets 22.4 miles per gallon, and his gas tank can hold 17.82 gallons of gas. How many miles can Gunnar travel if he uses all of the gas in the gas tank?

4. The principal of East High School wants to buy a new cover for the sand pit used in the long-jump competition. He measured the sand pit and found that the length is 29.2 feet and the width is 9.8 feet. What will the area of the new cover be?

Name _____   Date _____

Use estimation or fraction multiplication to determine if your answer is reasonable.

1.   Calculate the product. 78.93 × 32.45

2.   Paint costs $29.95 per gallon. Nikki needs 12.25 gallons to complete a painting project. How much will Nikki spend on paint? Remember to round to the nearest penny.

Solve each problem. Remember to round to the nearest penny when necessary.

1. Calculate the product. 64.13×19.39

   $64.13 \times 19.39 = 1{,}243.4807$

   > I know decimal multiplication is similar to whole number multiplication, but I have to determine where the decimal point is placed in the product. I can estimate the factors and determine the estimated product. $60 \times 20 = 1{,}200$. In the actual answer, the decimal point must be in a place where the product is close to 1,200. I can multiply using the algorithm and then place the decimal point after the ones place. 1,243.4807 is close to 1,200, so I know my answer is reasonable, and I correctly placed the decimal point.

   > I can also count the decimal digits in the first factor (2) and the decimal digits in the second factor (2) and add them together. $2 + 2 = 4$, so the product will have four decimal digits.

2. Every weekend, Talia visits the farmer's market and buys 5 grapefruits for $0.61 each and a loaf of banana bread for $6.99. How much does Talia spend at the farmer's market every weekend?

   $\$6.99 + (5 \times \$0.61) = \$10.04$

   > $5 \times \$0.60$ is $3.00, so Talia spends about $3.00 each weekend on grapefruit. I can add the cost of the bread, which is about $7, so Talia spends about $10 every weekend at the farmer's market. This estimated product could help me determine the correct placement of the decimal point. I can find the value of the expression in the parentheses first. $5 \times \$0.61 = \$3.05$. Now I can add both parts of the number sentence. $\$3.05 + \$6.99 = \$10.04$. This answer is close to the estimated answer of $10, so I know my answer is reasonable and the decimal point is in the correct place.

Solve each problem. Remember to round to the nearest penny when necessary.

1. Calculate the product. 45.67×32.58

2. Deprina buys a large cup of coffee for $4.70 on her way to work every day. If there are 24 workdays in the month, how much does Deprina spend on coffee throughout the entire month?

3. Krego earns $2,456.75 every month. He also earns an extra $4.75 every time he sells a new gym membership. Last month, Krego sold 32 new gym memberships. How much money did Krego earn last month?

4. Kendra just bought a new house and needs to buy new sod for her backyard. If the dimensions of her yard are 24.6 feet by 14.8 feet, what is the area of her yard?

Discussion

Divide 150 by 30.

Exercises 1–5

Round to estimate the quotient. Then, compute the quotient using a calculator, and compare the estimation to the quotient.

1. 2,970 ÷ 11

    a. Round to a one-digit arithmetic fact. Estimate the quotient.

    b. Use a calculator to find the quotient. Compare the quotient to the estimate.

2. $4,752 \div 12$

    a. Round to a one-digit arithmetic fact. Estimate the quotient.

    b. Use a calculator to find the quotient. Compare the quotient to the estimate.

3. $11,647 \div 19$

    a. Round to a one-digit arithmetic fact. Estimate the quotient.

    b. Use a calculator to find the quotient. Compare the quotient to the estimate.

Lesson 12: Estimating Digits in a Quotient

EUREKA MATH

4. 40,644 ÷ 18

    a.   Round to a one-digit arithmetic fact. Estimate the quotient.

    b.   Use a calculator to find the quotient. Compare the quotient to the estimate.

5. 49,170 ÷ 15

    a.   Round to a one-digit arithmetic fact. Estimate the quotient.

    b.   Use a calculator to find the quotient. Compare the quotient to the estimate.

Example 3: **Extend Estimation and Place Value to the Division Algorithm**

Estimate and apply the division algorithm to evaluate the expression $918 \div 27$.

     Lesson 12:   Estimating Digits in a Quotient

Name _____   Date _____

Round to estimate the quotient. Then, compute the quotient using a calculator, and compare the estimation to the quotient.

1. $4,732 \div 13$

2. $22,752 \div 16$

Round to estimate the quotient. Then, compute the quotient using a calculator, and compare the estimate to the quotient.

1. $891 \div 11 =$

   Estimate: $900 \div 10 = 90$

   Quotient: $891 \div 11 = 81$

   Comparison: Since the divisor is very close to a multiple of 10, the quotient is very close to the estimate.

   > I can round 11 to 10. I can round 891 to 900 since 900 is a multiple of 10. I can also choose to round 891 to 890 since it's a multiple of 10, and it would be easy to divide 890 by 10 also.

2. $13,616 \div 16 =$

   Estimate: $14,000 \div 20 = 700$

   Quotient: $13,616 \div 16 = 851$

   Comparison: The divisor is not close to a multiple of 10, so the quotient is not nearly as close to the estimate as when divisors are closer to a multiple of 10.

   > I can round the divisor to 20 because it's easier to divide by a divisor that is a multiple of 10, and 16 is closer to 20 than 10. I can round the dividend, 13,616, to 14,000 since it is closer to 14,000 than 13,000.

   > Divisors with digits 4, 5, and 6 in the ones place have less accurate estimates. Because the divisor in the problem is not very close to a multiple of 10, the estimate is not very close to the quotient.

Round to estimate the quotient. Then, compute the quotient using a calculator, and compare the estimate to the quotient.

1. $715 \div 11$

2. $7,884 \div 12$

3. $9,646 \div 13$

4. $11,942 \div 14$

5. $48,825 \div 15$

6. $135,296 \div 16$

7. $199,988 \div 17$

8. $116,478 \div 18$

9. $99,066 \div 19$

10. $181,800 \div 20$

Example 1

Divide 70,072 ÷ 19.

   a. Estimate:

   b. Create a table to show the multiples of 19.

Multiples of 19

c. Use the algorithm to divide $70,072 \div 19$. Check your work.

$$19 \overline{)\ 7\quad 0\quad 0\quad 7\quad 2\ }$$

Example 2

Divide $14,175 \div 315$.

   a. Estimate:

   b. Use the algorithm to divide $14,175 \div 315$. Check your work.

Exercises 1–5

For each exercise,

    a.   Estimate.

    b.   Divide using the algorithm, explaining your work using place value.

1. $484,692 \div 78$

    a.   Estimate:

    b.

2. $281,886 \div 33$

    a.   Estimate:

    b.

3. 2,295,517 ÷ 37

    a. Estimate:

    b.

4. 952,448 ÷ 112

    a. Estimate:

    b.

Lesson 13:   Dividing Multi-Digit Numbers Using the Algorithm

EUREKA
MATH®

5. 1,823,535 ÷ 245

    a. Estimate:

    b.

Name _____   Date _____

Divide using the division algorithm: $392{,}196 \div 87$.

Divide using the division algorithm.

1,332 ÷ 18

The quotient is 74.

Multiples of 18
$1 \times 18 = 18$
$2 \times 18 = 36$
$3 \times 18 = 54$
$4 \times 18 = 72$
$5 \times 18 = 90$
$6 \times 18 = 108$
$7 \times 18 = 126$
$8 \times 18 = 144$
$9 \times 18 = 162$

> I can use the tables of multiples to see that I can divide 133 tens into about 70 groups of 18.

$$
\begin{array}{r}
7\;4 \\
18\;\overline{)\;1,\;3\;3\;2} \\
5 \\
-\;1\;2\;6 \\
\hline
7\;2 \\
3 \\
-\;7\;2 \\
\hline
0
\end{array}
$$

> Now I can regroup and determine how many times 18 divides into 72. I see from my table of multiples that $18 \times 4 = 72$. I multiply 4 ones × 8 ones and get 3 tens and 2 ones. I record the 3 in the tens place and the 2 in the ones place.
> 4 ones × 10 ones is 4 tens, plus the 3 tens in the tens place is 7 tens. So the quotient is 74.

> I can round the dividend to 140 tens and the divisor to 2 tens. $1,400 \div 20 = 70$. Using this estimation and the table of multiples, 18 divides into 133 around 7 times, so I record 7 in the tens place.
> 7×8 ones is 5 tens and 6 ones, so I record the 5 in the tens place and the 6 in the ones place. 7×10 is 70; but when I add the 5 tens (or 50), I get 120, so I record the 1 in the hundreds place and the 2 in the tens place. I remember to cross out the 5. $133 - 126 = 7$.

Divide using the division algorithm.

1. $1,634 \div 19$

2. $2,450 \div 25$

3. $22,274 \div 37$

4. $21,361 \div 41$

5. $34,874 \div 53$

6. $50,902 \div 62$

7. $70,434 \div 78$

8. $91,047 \div 89$

9. $115,785 \div 93$

10. $207,968 \div 97$

11. $7,735 \div 119$

12. $21,948 \div 354$

13. $72,372 \div 111$

14. $74,152 \div 124$

15. $182,727 \div 257$

16. $396,256 \div 488$

17. $730,730 \div 715$

18. $1,434,342 \div 923$

19. $1,775,296 \div 32$

20. $1,144,932 \div 12$

Opening Exercise

Divide $\frac{1}{2} \div \frac{1}{10}$. Use a tape diagram to support your reasoning.

Relate the model to the invert and multiply rule.

Example 1

Evaluate the expression. Use a tape diagram to support your answer.

$0.5 \div 0.1$

Rewrite $0.5 \div 0.1$ as a fraction.

Express the divisor as a whole number.

Exercises 1–3

Convert the decimal division expressions to fractional division expressions in order to create whole number divisors. You do not need to find the quotients. Explain the movement of the decimal point. The first exercise has been completed for you.

1. $18.6 \div 2.3$

    $$\frac{18.6}{2.3} \times \frac{10}{10} = \frac{186}{23}$$

    $186 \div 23$

    I multiplied both the dividend and the divisor by ten, or by one power of ten, so each decimal point moved one place to the right because they grew larger by ten.

2. $14.04 \div 4.68$

3. $0.162 \div 0.036$

Example 2

Evaluate the expression. First, convert the decimal division expression to a fractional division expression in order to create a whole number divisor.

$25.2 \div 0.72$

Use the division algorithm to find the quotient.

Exercises 4–7

Convert the decimal division expressions to fractional division expressions in order to create whole number divisors. Compute the quotients using the division algorithm. Check your work with a calculator.

4. $2,000 \div 3.2$

5. $3,581.9 \div 4.9$

Lesson 14:   The Division Algorithm—Converting Decimal Division
             into Whole Number Division Using Fractions

EUREKA
MATH®

6. $893.76 \div 0.21$

7. $6.194 \div 0.326$

EUREKA
MATH®

Lesson 14:   The Division Algorithm—Converting Decimal Division
into Whole Number Division Using Fractions

149

© 2019 Great Minds®. eureka-math.org

Example 3

A plane travels 3,625.26 miles in 6.9 hours. What is the plane's unit rate?

Represent this situation with a fraction.

Represent this situation using the same units.

Estimate the quotient.

Express the divisor as a whole number.

Use the division algorithm to find the quotient.

Use multiplication to check your work.

Lesson 14:   The Division Algorithm—Converting Decimal Division
               into Whole Number Division Using Fractions

EUREKA
MATH

Name _____ Date _____

Estimate quotients. Convert decimal division expressions to fractional division expressions to create whole number divisors. Compute the quotient using the division algorithm. Check your work with a calculator and your estimate.

1. Lisa purchased almonds for $3.50 per pound. She spent a total of $24.50. How many pounds of almonds did she purchase?

2. Divide: $125.01 \div 5.4$.

1. Convert decimal division expressions to fractional division expressions to create whole number divisors.

   $624.12 \div 0.8$

   $$\frac{624.12}{0.8} \times \frac{10}{10} = \frac{6,241.2}{8}$$

   > To convert the divisor, 0.8, to a whole number, I can multiply by 10. I must also multiply the dividend by 10 for equality.

2. Estimate quotients. Convert decimal division expressions to fractional division expressions to create whole number divisors. Compute the quotients using the division algorithm. Check your work with a calculator and your estimates.

   Nicky purchased several notepads for $3.70 each. She spent a total of $29.60. How many notepads did she buy?

   $$\frac{29.60}{3.7} \times \frac{10}{10} = \frac{296}{37}$$

   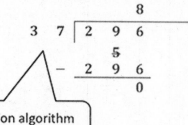

   > I can convert the divisor to a whole number.

   > I can use the division algorithm to find the quotient.

   Estimate: $32 \div 4 = 8$

   Nicky purchased 8 notebooks, so the quotient is the same as the estimate.

EUREKA
MATH

© 2019 Great Minds®. eureka-math.org

Convert decimal division expressions to fractional division expressions to create whole number divisors.

1. $35.7 \div 0.07$

2. $486.12 \div 0.6$

3. $3.43 \div 0.035$

4. $5,418.54 \div 0.009$

5. $812.5 \div 1.25$

6. $17.343 \div 36.9$

Estimate quotients. Convert decimal division expressions to fractional division expressions to create whole number divisors. Compute the quotients using the division algorithm. Check your work with a calculator and your estimates.

7. Norman purchased 3.5 lb. of his favorite mixture of dried fruits to use in a trail mix. The total cost was $16.87. How much does the fruit cost per pound?

8. Divide: $994.14 \div 18.9$

9. Daryl spent $4.68 on each pound of trail mix. He spent a total of $14.04. How many pounds of trail mix did he purchase?

10. Mamie saved $161.25. This is 25% of the amount she needs to save. How much money does Mamie need to save?

11. Kareem purchased several packs of gum to place in gift baskets for $1.26 each. He spent a total of $8.82. How many packs of gum did he buy?

12. Jerod is making candles from beeswax. He has 132.72 ounces of beeswax. If each candle uses 8.4 ounces of beeswax, how many candles can he make? Will there be any wax left over?

13. There are 20.5 cups of batter in the bowl. This represents 0.4 of the entire amount of batter needed for a recipe. How many cups of batter are needed?

14. Divide: $159.12 \div 6.8$

15. Divide: $167.67 \div 8.1$

Opening Exercise

Use mental math to evaluate the numeric expressions.

   a. $99 + 44$

   b. $86 - 39$

   c. 50×14

   d. $180 \div 5$

Example 1: Use Mental Math to Find Quotients

Use mental math to evaluate $105 \div 35$.

Exercises 1–4

Use mental math techniques to evaluate the expressions.

1. $770 \div 14$

2. $1,005 \div 5$

3. $1,500 \div 8$

4. $1,260 \div 5$

EUREKA
MATH®

Example 2: Mental Math and Division of Decimals

Evaluate the expression $175 \div 3.5$ using mental math techniques.

Exercises 5–7

Use mental math techniques to evaluate the expressions.

5. $25 \div 6.25$

6. $6.3 \div 1.5$

7. $425 \div 2.5$

Example 3: Mental Math and the Division Algorithm

Evaluate the expression $4{,}564 \div 3.5$ using mental math techniques and the division algorithm.

EUREKA
MATH

Example 4: Mental Math and Reasonable Work

Shelly was given this number sentence and was asked to place the decimal point correctly in the quotient.

$$55.6875 \div 6.75 = 0.825$$

Do you agree with Shelly?

Divide to prove your answer is correct.

EUREKA
MATH

Lesson 15:   The Division Algorithm–Converting Decimal Division
into Whole Number Division Using Mental Math

© 2019 Great Minds®. eureka-math.org

161

Name _____   Date _____

Evaluate the expression using mental math techniques and the division algorithm. Explain your reasoning.

$$18.75 \div 2.5$$

1. Use mental math, estimation, and the division algorithm to evaluate the expressions.

    $405 \div 4.5$

    Mental Math: $810 \div 9 = 90$

    > I can multiply the dividend and the divisor by 2 to get a whole number divisor. This also creates a whole number that easily divides into the dividend that has been doubled.

    Estimate: $400 \div 4 = 100$

    Algorithm:

    ```
              9  0
      4 5 | 4, 0 5 0
            4
          - 4 0 5
              0
    ```

    > Since $45 \times 10 = 450$ and 450 is larger than 405, I can try 45×9. I multiplied 9 tens by 5 ones and got 4 hundreds and 5 tens. I multiplied 9 tens by 4 tens and got 3 thousands 6 hundreds. I added the 4 hundreds to the 6 hundreds, and that is another thousand, so the total is 4 thousands, 5 tens. Now I can subtract, and the difference is 0. The quotient is 90.

2. Place the decimal point in the correct place to make the number sentence true.

    $65.5872 \div 6.1 = 10752$

    $65.5872 \div 6.1 = 10.752$

    > I can round the dividend to 66 and the divisor to 6. The quotient is 11. The decimal point in the quotient is placed after the ones place. The decimal point is in the correct place because 10.752 is close to 11, my estimated quotient.

Use mental math, estimation, and the division algorithm to evaluate the expressions.

1. $118.4 \div 6.4$

2. $314.944 \div 3.7$

3. $1,840.5072 \div 23.56$

4. $325 \div 2.5$

5. $196 \div 3.5$

6. $405 \div 4.5$

7. $3,437.5 \div 5.5$

8. $393.75 \div 5.25$

9. $2,625 \div 6.25$

10. $231 \div 8.25$

11. $92 \div 5.75$

12. $196 \div 12.25$

13. $117 \div 6.5$

14. $936 \div 9.75$

15. $305 \div 12.2$

Place the decimal point in the correct place to make the number sentence true.

16. $83.375 \div 2.3 = 3,625$

17. $183.575 \div 5,245 = 3.5$

18. $326,025 \div 9.45 = 3.45$

19. $449.5 \div 725 = 6.2$

20. $446,642 \div 85.4 = 52.3$

Opening Exercise

   a. What is an even number?

   b. List some examples of even numbers.

   c. What is an odd number?

   d. List some examples of odd numbers.

What happens when we add two even numbers? Do we always get an even number?

Exercises 1–3

1. Why is the sum of two even numbers even?

    a. Think of the problem $12 + 14$. Draw dots to represent each number.

    b. Circle pairs of dots to determine if any of the dots are left over.

    c. Is this true every time two even numbers are added together? Why or why not?

2. Why is the sum of two odd numbers even?

    a. Think of the problem $11 + 15$. Draw dots to represent each number.

    b. Circle pairs of dots to determine if any of the dots are left over.

    c. Is this true every time two odd numbers are added together? Why or why not?

3. Why is the sum of an even number and an odd number odd?

   a. Think of the problem $14 + 11$. Draw dots to represent each number.

   b. Circle pairs of dots to determine if any of the dots are left over.

   c. Is this true every time an even number and an odd number are added together? Why or why not?

   d. What if the first addend is odd and the second is even? Is the sum still odd? Why or why not? For example, if we had $11 + 14$, would the sum be odd?

Let's sum it up:

-
-
-

Exploratory Challenge/Exercises 4–6

4. The product of two even numbers is even.

5. The product of two odd numbers is odd.

6. The product of an even number and an odd number is even.

EUREKA
MATH

Lesson Summary

Adding:

- The sum of two even numbers is even.
- The sum of two odd numbers is even.
- The sum of an even number and an odd number is odd.

Multiplying:

- The product of two even numbers is even.
- The product of two odd numbers is odd.
- The product of an even number and an odd number is even.

Name _____ Date _____

Determine whether each sum or product is even or odd. Explain your reasoning.

1. $56,426 + 17,895$

2. $317,362 \times 129,324$

3. $10,481 + 4,569$

4. $32,457 \times 12,781$

5. Show or explain why $12 + 13 + 14 + 15 + 16$ results in an even sum.

Lesson Notes

Adding:

- The sum of two even numbers is even.
- The sum of two odd numbers is even.
- The sum of an even number and an odd number is odd.

Multiplying:

- The product of two even numbers is even.
- The product of two odd numbers is odd.
- The product of an even number and an odd number is even.

1. When solving, tell whether the sum is even or odd. Explain your reasoning.

   $951 + 244$

   > When I add these two numbers, the odd number will have a dot remaining after I circle pairs of dots. The even number will not have any dots remaining after I circle the pairs of dots, so the one remaining dot from the odd number will not be able to join with another dot to make a pair. The sum is odd.

   The sum is odd because the sum of an odd number and an even number is odd.

2. When solving, tell whether the product is even or odd. Explain your reasoning.

   $2,422 \times 346$

   > In this problem, I have 2,422 groups of 346, so I have an even number of groups of 346. When I add the addends (346) two at a time, the sum is always even because there are no dots remaining after I circle all the pairs.

   The product is even because the product of two even numbers is even.

Without solving, tell whether each sum or product is even or odd. Explain your reasoning.

1. 346 + 721

2. 4,690 × 141

3. 1,462,891 × 745,629

4. 425,922 + 32,481,064

5. 32 + 45 + 67 + 91 + 34 + 56

Opening Exercise

Below is a list of 10 numbers. Place each number in the circle(s) that is a factor of the number. Some numbers can be placed in more than one circle. For example, if 32 were on the list, it would be placed in the circles with 2, 4, and 8 because they are all factors of 32.

24; 36; 80; 115; 214; 360; 975; 4,678; 29,785; 414,940

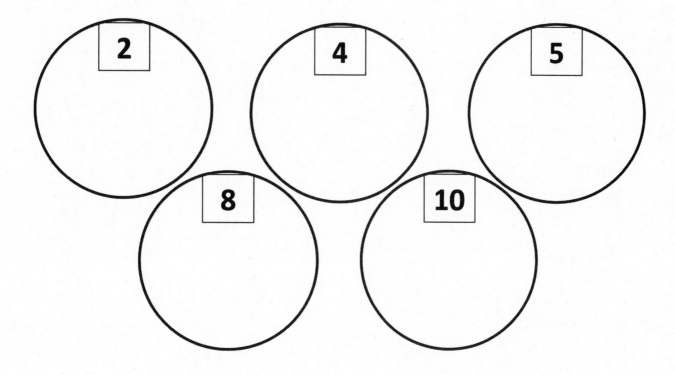

EUREKA MATH®

Discussion

- Divisibility rule for 2:

- Divisibility rule for 4:

- Divisibility rule for 5:

- Divisibility rule for 8:

- Divisibility rule for 10:

- Decimal numbers with fraction parts do not follow the divisibility tests.

- Divisibility rule for 3:

- Divisibility rule for 9:

Example 1

This example shows how to apply the two new divisibility rules we just discussed.

Explain why 378 is divisible by 3 and 9.

a. Expand 378.

b. Decompose the expression to factor by 9.

c. Factor the 9.

d. What is the sum of the three digits?

e. Is 18 divisble by 9?

f. Is the number 378 divisible by 9? Why or why not?

g. Is the number 378 divisible by 3? Why or why not?

Example 2

Is 3,822 divisible by 3 or 9? Why or why not?

Exercises 1–5

Circle ALL the numbers that are factors of the given number. Complete any necessary work in the space provided.

1. 2,838 is divisible by

    3

    9

    4

    Explain your reasoning for your choice(s).

2. 34,515 is divisible by

    3

    9

    5

    Explain your reasoning for your choice(s).

3. 10,534,341 is divisible by

    3

    9

    2

    Explain your reasoning for your choice(s).

EUREKA
MATH®

4.   4,320 is divisible by

3

9

10

Explain your reasoning for your choice(s).

5.   6,240 is divisible by

3

9

8

Explain your reasoning for your choice(s).

Lesson Summary

To determine if a number is divisible by 3 or 9:

- Calculate the sum of the digits.
- If the sum of the digits is divisible by 3, the entire number is divisible by 3.
- If the sum of the digits is divisible by 9, the entire number is divisible by 9.

Note: If a number is divisible by 9, the number is also divisible by 3.

Lesson 17: Divisibility Tests for 3 and 9

EUREKA MATH

Name _____ Date _____

1. Is 26,341 divisible by 3? If it is, write the number as the product of 3 and another factor. If not, explain.

2. Is 8,397 divisible by 9? If it is, write the number as the product of 9 and another factor. If not, explain.

3. Explain why 186,426 is divisible by both 3 and 9.

1. Is 5,641 divisible by both 3 and 9? Why or why not?

   The number 5,641 *is not divisible by 3 and 9 because the sum of the digits is 16, which is not divisible by 3 or 9.*

   > I can find the sum of the digits by adding $5 + 6 + 4 + 1$. The sum is 16.

   > If the sum of the digits is 15, the number would be divisible by 3 but not 9 since 15 is divisible by 3 but not 9. If the sum of the digits is 27, the number would be divisible by 3 and 9 since 27 is a multiple of 3 and 9.

2. Circle all the factors of 71,820 from the list below.

   ② ③ ④ ⑤ 8 ⑨ ⑩

   > 71,820 is an even number, so it is divisible by 2. When I added $7 + 1 + 8 + 2 + 0$, the sum is 18, which is divisible by 3 and 9, so the entire number is divisible by 3 and 9. The last 2 digits, 20, are divisible by 4, so the entire number is divisible by 4. The number ends in a 0, so the entire number is also divisible by 5 and 10.

3. Write a 3-digit number that is divisible by both 3 and 4. Explain how you know this number is divisible by 3 and 4.

   324 is a 3-digit number that is divisible by 3 and 4 because the sum of the digits is divisible by 3, and the last two digits are divisible by 4.

   > I know the number has to have three digits, and since it is divisible by 4, the last 2 digits have to be divisible by 4. So, I can write a number that ends in 24 since 24 is divisible by 4. Since $2 + 4$ is 6, and I need to make a 3-digit number, 3 more is 9, which is divisible by 3. So my number is 324.

1. Is 32,643 divisible by both 3 and 9? Why or why not?

2. Circle all the factors of 424,380 from the list below.

   2   3   4   5   8   9   10

3. Circle all the factors of 322,875 from the list below.

   2   3   4   5   8   9   10

4. Write a 3-digit number that is divisible by both 3 and 4. Explain how you know this number is divisible by 3 and 4.

5. Write a 4-digit number that is divisible by both 5 and 9. Explain how you know this number is divisible by 5 and 9.

Opening

The *greatest common factor* of two whole numbers (not both zero) is the greatest whole number that is a factor of each number. The greatest common factor of two whole numbers a and b is denoted by GCF (a, b).

The *least common multiple* of two whole numbers is the smallest whole number greater than zero that is a multiple of each number. The least common multiple of two whole numbers a and b is denoted by LCM (a, b).

Example 1: Greatest Common Factor

Find the greatest common factor of 12 and 18.

- Listing these factor pairs in order helps ensure that no common factors are missed. Start with 1 multiplied by the number.
- Circle all factors that appear on both lists.
- Place a triangle around the greatest of these common factors.

GCF (12, 18)

12

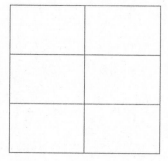

18

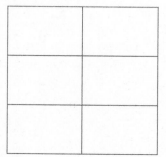

Example 2: Least Common Multiple

Find the least common multiple of 12 and 18.

LCM (12, 18)

Write the first 10 multiples of 12.

Write the first 10 multiples of 18.

Circle the multiples that appear on both lists.

Put a rectangle around the least of these common multiples.

Exercises

Station 1: Factors and GCF

Choose one of these problems that has not yet been solved. Solve it together on your student page. Then, use your marker to copy your work neatly on the chart paper. Use your marker to cross out your choice so that the next group solves a different problem.

GCF (30, 50)

GCF (30, 45)

GCF (45, 60)

GCF (42, 70)

GCF (96, 144)

Lesson 18:   Least Common Multiple and Greatest Common Factor

EUREKA MATH

Next, choose one of these problems that has not yet been solved:

a. There are 18 girls and 24 boys who want to participate in a Trivia Challenge. If each team must have the same ratio of girls and boys, what is the greatest number of teams that can enter? Find how many boys and girls each team would have.

b. Ski Club members are preparing identical welcome kits for new skiers. The Ski Club has 60 hand-warmer packets and 48 foot-warmer packets. Find the greatest number of identical kits they can prepare using all of the hand-warmer and foot-warmer packets. How many hand-warmer packets and foot-warmer packets would each welcome kit have?

c. There are 435 representatives and 100 senators serving in the United States Congress. How many identical groups with the same numbers of representatives and senators could be formed from all of Congress if we want the largest groups possible? How many representatives and senators would be in each group?

d. Is the GCF of a pair of numbers ever equal to one of the numbers? Explain with an example.

e. Is the GCF of a pair of numbers ever greater than both numbers? Explain with an example.

Station 2: Multiples and LCM

Choose one of these problems that has not yet been solved. Solve it together on your student page. Then, use your marker to copy your work neatly on the chart paper. Use your marker to cross out your choice so that the next group solves a different problem.

LCM (9, 12)

LCM (8, 18)

LCM (4, 30)

LCM (12, 30)

LCM (20, 50)

Next, choose one of these problems that has not yet been solved. Solve it together on your student page. Then, use your marker to copy your work neatly on this chart paper and to cross out your choice so that the next group solves a different problem.

a. Hot dogs come packed 10 in a package. Hot dog buns come packed 8 in a package. If we want one hot dog for each bun for a picnic with none left over, what is the least amount of each we need to buy? How many packages of each item would we have to buy?

b. Starting at 6:00 a.m., a bus stops at my street corner every 15 minutes. Also starting at 6:00 a.m., a taxi cab comes by every 12 minutes. What is the next time both a bus and a taxi are at the corner at the same time?

c. Two gears in a machine are aligned by a mark drawn from the center of one gear to the center of the other. If the first gear has 24 teeth, and the second gear has 40 teeth, how many revolutions of the first gear are needed until the marks line up again?

Lesson 18: Least Common Multiple and Greatest Common Factor

EUREKA MATH®

d. Is the LCM of a pair of numbers ever equal to one of the numbers? Explain with an example.

e. Is the LCM of a pair of numbers ever less than both numbers? Explain with an example.

Station 3: Using Prime Factors to Determine GCF

Choose one of these problems that has not yet been solved. Solve it together on your student page. Then, use your marker to copy your work neatly on the chart paper and to cross out your choice so that the next group solves a different problem.

GCF $(30, 50)$

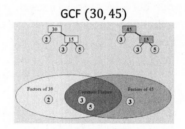

GCF $(30, 45)$

GCF $(45, 60)$

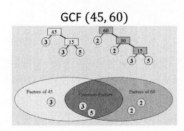

GCF $(42, 70)$

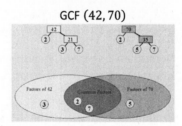

GCF $(96, 144)$

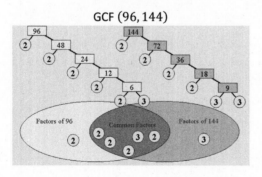

Next, choose one of these problems that has not yet been solved:

   a.   Would you rather find all the factors of a number or find all the prime factors of a number? Why?

   b.   Find the GCF of your original pair of numbers.

   c.   Is the product of your LCM and GCF less than, greater than, or equal to the product of your numbers?

   d.   Glenn's favorite number is very special because it reminds him of the day his daughter, Sarah, was born. The factors of this number do not repeat, and all the prime numbers are less than 12. What is Glenn's number? When was Sarah born?

Station 4: Applying Factors to the Distributive Property

Choose one of these problems that has not yet been solved. Solve it together on your student page. Then, use your marker to copy your work neatly on the chart paper and to cross out your choice so that the next group solves a different problem.

Find the GCF from the two numbers, and rewrite the sum using the distributive property.

1. $12 + 18 =$

2. $42 + 14 =$

3. $36 + 27 =$

4. $16 + 72 =$

5. $44 + 33 =$

Lesson 18:   Least Common Multiple and Greatest Common Factor

EUREKA MATH®

Next, add another example to one of these two statements applying factors to the distributive property.

Choose any numbers for n, a, and b.

$$n(a) + n(b) = n(a + b)$$

$$n(a) - n(b) = n(a - b)$$

Name _____   Date _____

1. Find the LCM and GCF of 12 and 15.

2. Write two numbers, neither of which is 8, whose GCF is 8.

3. Write two numbers, neither of which is 28, whose LCM is 28.

Rate each of the stations you visited today. Use this scale:

3—Easy—I've got it; I don't need any help.

2—Medium—I need more practice, but I understand some of it.

1—Hard—I'm not getting this yet.

Complete the following chart:

Station	Rating (3, 2, 1)	Comment to the Teacher
Station 1: Factors and GCF		
Station 2: Multiples and LCM		
Station 3: Using Prime Factors for GCF		
Station 4: Applying Factors to the Distributive Property		

Factors and GCF

1. The Knitting Club members are preparing identical welcome kits for new members. The Knitting Club has 45 spools of yarn and 75 knitting needles. Find the greatest number of identical kits they can prepare using all of the yarn and knitting needles. How many spools of yarn and knitting needles would each welcome kit have?

Factors of 45:

1, 3, 5, 9, 15, 45

Common Factors:

1, 3, 5, 15

Factors of 75:

1, 3, 5, 15, 25, 75

GCF:

15

GCF (45, 75) *is* 15. *There would be* 15 *identical kits. Each kit will have* 3 *spools of yarn and* 5 *knitting needles.*

> I can find the GCF of 45 and 75 by listing the factors of each number and the common factors and by identifying the greatest factor both numbers have in common (the GCF).

> Since there are 15 kits and a total of 45 spools of yarn, 45 ÷ 15 = 3, so each kit will have 3 spools of yarn.

> Since there are 15 kits and a total of 75 knitting needles, 75 ÷ 15 = 5, so each kit will have 5 knitting needles.

Multiples and LCM

2. Madison has two plants. She waters the spider plant every 4 days and the cactus every 6 days. She watered both plants on November 30. What is the next day that she will water both plants?

The LCM of 4 *and* 6 *is* 12, *so she will water both plants on December* 12.

> I can also list the multiples of each number until I find one that both have in common. I will list the first five multiples of each number although I can stop whenever I identify a common multiple.
> Multiples of 4: 4, 8, ⑫ 16, 20
> Multiples of 6: 6, ⑫ 18, 24, 30

Lesson 18: Least Common Multiple and Greatest Common Factor

203

© 2019 Great Minds®. eureka-math.org

Using Prime Factors to Determine GCF

3. Use prime factors to find the greatest common factor of the following pairs of numbers.

   GCF (18, 27)

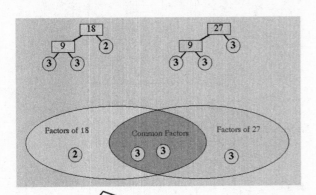

I can find the prime factors of 18 and 27 by decomposing each number using the factor tree.

I can multiply the shared factors to find the greatest common factor (GCF).
$3 \times 3 = 9$.

I can use the Venn diagram to compare and organize the factors. I can put the common factors in the middle section of the Venn diagram and the unique factors in the left and right parts.

GCF (18, 27) = 9

Applying Factors to the Distributive Property

4. Find the GCF from the two numbers, and rewrite the sum using the distributive property.

   $16 + 40 =$

   GCF (16, 40) = 8

   $16 + 40 = 8(2) + 8(5) = 8(2 + 5) = 8(7) = 56$

I can determine the GCF of 16 and 40, which is 8. I can rewrite 16 and 40 by factoring out the GCF. $8 \times 2 = 16$, and $8 \times 5 = 40$.

EUREKA
MATH

Complete the remaining stations from class.

Opening Exercise

Euclid's algorithm is used to find the greatest common factor (GCF) of two whole numbers.

1. Divide the larger of the two numbers by the smaller one.

2. If there is a remainder, divide it into the divisor.

3. Continue dividing the last divisor by the last remainder until the remainder is zero.

4. The final divisor is the GCF of the original pair of numbers.

$383 \div 4 =$

$432 \div 12 =$

$403 \div 13 =$

Example 1: Euclid's Algorithm Conceptualized

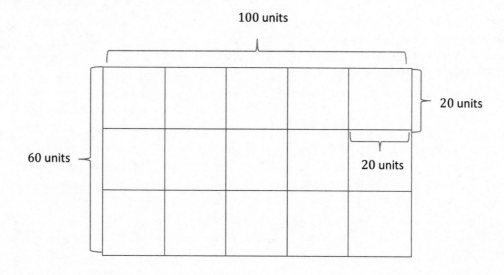

Example 2: Lesson 18 Classwork Revisited

a. Let's apply Euclid's algorithm to some of the problems from our last lesson.

    i.   What is the GCF of 30 and 50?

    ii.   Using Euclid's algorithm, we follow the steps that are listed in the Opening Exercise.

b. Apply Euclid's algorithm to find the GCF (30, 45).

Example 3: Larger Numbers

        GCF (96, 144)   GCF (660, 840)

Example 4: Area Problems

The greatest common factor has many uses. Among them, the GCF lets us find out the maximum size of squares that cover a rectangle. When we solve problems like this, we cannot have any gaps or any overlapping squares. Of course, the maximum size squares will be the minimum number of squares needed.

A rectangular computer table measures 30 inches by 50 inches. We need to cover it with square tiles. What is the side length of the largest square tile we can use to completely cover the table without overlap or gaps?

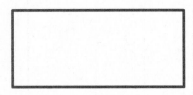

a. If we use squares that are 10 by 10, how many do we need?

b. If this were a giant chunk of cheese in a factory, would it change the thinking or the calculations we just did?

c. How many 10 inch × 10 inch squares of cheese could be cut from the giant 30 inch × 50 inch slab?

Name _____   Date _____

Use Euclid's algorithm to find the greatest common factor of 45 and 75.

1. Use Euclid's algorithm to find the greatest common factor of the following pairs of numbers.

   GCF (16, 158)

   GCF (16, 158) = 2

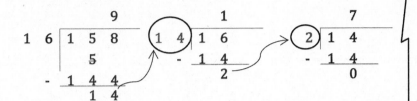

> I can divide 158 by 16 since 158 is the larger of the two numbers. There is a remainder of 14, so I can divide the divisor, 16, by the remainder, 14. There is another remainder of 2, so I can divide the divisor, 14, by the remainder again. $14 \div 2 = 7$, and there is no remainder. Since 2 is the final divisor, 2 is the GCF of the original pair of numbers, 16 and 158.

2. Kristen and Alen are planning a party for their son's birthday. They order a rectangular cake that measures 12 inches by 18 inches.

   a. All pieces of the cake must be square with none left over. What is the side length of the largest square pieces into which Kristen and Alen can cut the cake?

   GCF (12, 18) = 6

   They can cut the cake into 6 inch by 6 inch squares.

```
      1              2
1 2 | 1 8   →   6 | 1 2
    - 1 2         - 1 2
      6             0
```

> I can use Euclid's algorithm to find the greatest common factor of 12 and 18.

   b. How many pieces of this size can be cut?

   2 × 3 = 6

   Kristen and Alen can cut 6 pieces of cake.

> I can visualize the whole cake, which is 12 inches by 18 inches. Since the GCF is 6, I can cut the 12-inch side in half since $12 \div 6 = 2$. I can cut the 18-inch side into thirds since $18 \div 6 = 3$. Now I can multiply. $2 \times 3 = 6$, so Kristen and Alen can cut 6 pieces of cake.

1. Use Euclid's algorithm to find the greatest common factor of the following pairs of numbers:

   a. GCF (12, 78)

   b. GCF (18, 176)

2. Juanita and Samuel are planning a pizza party. They order a rectangular sheet pizza that measures 21 inches by 36 inches. They tell the pizza maker not to cut it because they want to cut it themselves.

   a. All pieces of pizza must be square with none left over. What is the side length of the largest square pieces into which Juanita and Samuel can cut the pizza?

   b. How many pieces of this size can be cut?

3. Shelly and Mickelle are making a quilt. They have a piece of fabric that measures 48 inches by 168 inches.

   a. All pieces of fabric must be square with none left over. What is the side length of the largest square pieces into which Shelly and Mickelle can cut the fabric?

   b. How many pieces of this size can Shelly and Mickelle cut?

Credits

Great Minds® has made every effort to obtain permission for the reprinting of all copyrighted material. If any owner of copyrighted material is not acknowledged herein, please contact Great Minds for proper acknowledgment in all future editions and reprints of this module.